S0-DMR-349

Environmental Health

Third world problems – first world preoccupations

Environmental Health

Third world problems – first world preoccupations

edited by Lorraine Mooney and Roger Bate

OXFORD AUCKLAND BOSTON JOHANNESBURG MELBOURNE NEW DELHI

Butterworth-Heinemann
Linacre House, Jordan Hill, Oxford OX2 8DP
225 Wildwood Avenue, Woburn, MA 01801-2041
A division of Reed Educational and Professional Publishing Ltd

A member of the Reed Elsevier plc group

First published 1999

British Library Cataloguing in Publication Data
A catalogue record for this book is available from the British Library

Library of Congress Cataloguing in Publication Data
A catalogue record for this book is available from the Library of Congress

ISBN 0 7506 4223 8

Printed in Great Britain by Biddles Ltd, Guildford and King's Lynn

Contents

Executive Summary vii
Biographies x
Introduction xiv

1 Malaria and the DDT story 1
Kelvin Kemm

2 Dirty water: Cholera in Peru 17
Enrique Ghersi and Hector Ñaupari

3 A sociology of health panics 47
Frank Furedi

4 Ecological risk: Actual and hypothetical 63
Kirill Kondratyev

5 Politics, policy, poisoning and food scares 85
Barrie Craven and Christine Johnson

6 Dietary nitrates pose no threat to human health 119
Jean Louis L'hirondel

7 Rachel's folly: The end of chlorine 129
Michelle Malkin and Michael Fumento

8 Organochlorines – natural and anthropogenic 161
Gordon Gribble

9 Nature's hormone factory 177
Jonathan Tolman

10 The saga of the falling sperm counts 215
James Le Fanu

11 Commentary 222
Lorraine Mooney

About ESEF 226
Authors' addresses 232
Index 235

Executive summary

This book underlines the differences in health priorities that exist around the world, and cautions against the export of inappropriate attitudes from developed to less developed countries.

American epidemiological studies in the mid-1980s showed an increased hypothetical cancer risk from trihalomethanes (THM), a by-product of the chlorination process. This hypothetical risk was extrapolated to population level: Seven hundred Americans a year were at risk. An anti-chlorine campaign, instigated by Greenpeace and other environmental groups, spread to Peru and elsewhere. This campaign contributed to the failure to adequately disinfect water supplies, and residual chlorine levels were far below the WHO minimum standard when cholera struck in 1991. An epidemic raged for five years, spreading to surrounding countries. In Peru alone, it killed more than 6000 people and put 625 000 into hospital.

Probably the most clear-cut illustration of inappropriate western attitudes having a baleful influence on the developing world is given by Kelvin Kemm. Dr Kemm is frustrated by the environmental campaign against the chlorine-based pesticide, DDT, which resulted in its being banned by many countries, including his own, South Africa. However, DDT is the developing world's most practical weapon against malaria. It is safe, cheap and specific. The claims made against it concerned its persistence in the environment and its contamination of the food chain. Malaria is increasing rapidly in sub-Saharan Africa, where 90 per cent of cases occur. Eradication campaigns in the 1960s that successfully removed malaria from North America and Europe, did so with the use of DDT. The campaign never attempted eradication in Africa, because it was too big a job. Funding for the programme was slashed in 1974, almost at the same time that DDT was banned. Dr Kemm is supporting recent moves in Zimbabwe to re-start spraying the insides of houses with DDT.

The health preoccupations of developed countries have a very different character. It is a modern paradox that in post-industrial societies, while the standard of health has never been better, life expectancy continues to grow, working and living environments never safer nor better monitored, yet there is a growing obsession with health and safety, even to the extent of a loss of judgement about risk assessment. Frank Furedi analyses the sociological factors which help to explain why people worry themselves to the point of illness.

Professor Kirill Kondratyev reviews the issues concerning environmental risk, which were raised in an earlier ESEF publication, 'What Risk? Science, Politics and Public Health'. These focus on the cancer risks from low doses of toxins found in the environment, and the measurement, portrayal and perception of these risks. The key point of interest is the existence of a threshold effect – that there is a level of exposure to a toxin which is too low to have any effect at all – the dose makes the poison. The issues discussed will be familiar as the subjects of environmental and health campaigns over the years – asbestos, dioxin, pesticides, environmental tobacco smoke, radiation – and in each case, when they are examined objectively and coolly, the risks are smaller than originally feared. Kondratyev insists that assessments of ecological risks must focus on identifying priorities and differentiating between real and hypothetical threats.

A similar case is made by Dr L'hirondel about the alleged effects of nitrate to health. The history of the concern surrounding nitrate fertiliser provides a very good illustration of the life cycle of a health scare. It goes back to the 1950s, when suspicion fell on nitrates in connection with methaemoglobinaemia in infants or blue-baby syndrome; then the attack shifted to a cancer threat from nitrosamines.

After many years of research, in this case covering two generations, as Dr L'hirondel is continuing his father's work on nitrates, all these suspicions have proved unfounded. He presents conclusive evidence that nitrates pose no threat to man, and calls for the repeal of directives limiting the nitrate content in drinking water, which are causing unnecessary expense.

Craven and Johnson examine the politics surrounding food scares and poisonings. Despite continually improving health throughout the 1990s, there was a four-fold increase in reported food poisoning cases. Dozens of new food regulations, designed to reduce risks from food, have not achieved the desired effect of protecting consumers, but they are imposing significant costs upon those whom the regulators are trying to protect: the producers and the consumers. While costs to the health services, the family and industry are increasing from the incidence of food poisoning, these are trivial by comparison with other causes of morbidity and mortality, such as back pain, which gets no media attention. The authors call for a balance to be struck between placing responsibility on the individual for their own actions and protecting the individual from the mistakes, incompetence or deceit of others. They find that too much emphasis is placed on attempting to protect the individual.

One of the most scientifically-illiterate environmental campaigns is the attempt to ban chlorine, or at least its man-made compounds. Malkin and Fumento trace the history of the campaign and discuss its implications. Amongst other things, they analyse dioxins, PCBs and plastics, and conclude that chlorine has been of enormous benefit to society. The eponymous Rachel Carson inadvertently sparked the campaign, and perhaps the whole environmental movement, when she wrote in 1962 of her fears that pesticides were damaging wildlife and could ultimately be harmful to man as well.

It is often repeated that man-made chemicals are somehow more hazardous than natural ones. Tolman disputes this in the case of environmental oestrogens – endocrine disrupters which upset the hormonal balance of animals. He says that naturally-occurring phytoestrogens produced by plants, are not only far more prolific than synthetic oestrogens, but also that their effects are about 40 million times greater, although neither have been found to pose a threat to human health.

Gribble conducts a fascinating review of natural organochlorines and their uses in many medicinal and industrial products. 'Organochlorines cannot be banned, any more than photosynthesis or gravity can be banned', he says. 'All toxic chemicals must be monitored and regulated intelligently with proper perspective, since nature will continue to produce organochlorines for its own purpose.'

Le Fanu picks up a similar thread in his review of the literature regarding the assertion that chemical pollutants which mimic the effect of the female hormone, oestrogen, known as xenoestrogens, are linked to low sperm counts, infertility and genital abnormalities. He finds that reliance has been placed on one piece of work – a meta-analysis of many separate studies, made over many years, in different countries – as evidence of a decline in sperm count. Flaws in this analysis make any assumption of falling sperm counts unsafe.

Editors' note

We draw attention to the individual summaries which appear at the beginning of each paper, and hope that these are useful in giving a quick understanding of the topics discussed in depth in the paper. The similarity of subject matter among some of the papers, particularly those regarding chlorine, cause some small overlap to occur, with similar topics being treated by different authors.

Acknowledgements

We would like to thank the Märit and Hans Rausing Charitable Foundation for their support of this book project.

LM & RB
Cambridge, November 1998

Biographies

Roger Bate

Roger Bate is the Director of the Environment Unit at the Institute of Economic Affairs in London, and a Director of ESEF. He is the author of several academic papers on science policy and economic issues, and has published numerous articles in newspapers such as the *Wall Street Journal* the *Financial Times, Sunday Times*. He is a columnist for the German media magazine, *Novo*, and *Economic Affairs*, and has appeared frequently on television and radio. He is a fellow of the Royal Society of Arts.

Barrie M. Craven

Dr Craven is Reader in economics at the University of Northumbria. He is a graduate of the University of Hull and the University of Newcastle upon Tyne. He has published in the field of monetary economics in the *Journal of Monetary Economics*, the *Manchester School* and the *European Journal of Finance*. More recently, he has researched public policy and health care issues. Current research is focused on the resource strategies associated with AIDS where results are published in several journals including the *Journal of Public Policy and Financial Accountability and Management*. He has taught at Curtin University, Western Australia, and at California Polytechnic in California, USA.

Frank Furedi

Dr Furedi teaches sociology at the University of Kent at Canterbury, England. His main area of research interest is the relationship between change and the perception of social problems. He is currently engaged on a major study of the sociology of political correctness. His latest book, 'Culture of Fear: Risk-taking and the Morality of Low Expectation' is published by Cassell, London, 1997.

Michael Fumento

Michael Fumento is a graduate of the University of Illinois School of Law and a former AIDS analyst and attorney for the US Commission of Civil Rights. He has also been a consultant on a National Institutes of Health AIDS project.

He was the 1994 Warren Brookes Fellow in Environmental Journalism at CEI. Prior to that, Mr Fumento was a reporter specialising in economics and science for *Investor's Business Daily*. He has written two books that have earned him a reputation as a meticulous researcher who pursues the truth: *Science Under Siege: Balancing Technology and the Environment*, and *The Myth of Heterosexual AIDS*. Mr Fumento received the American Council on Science and Health's 'Distinguished Science Journalist of 1993' award for *Science Under Siege*.

Enrique Ghersi

Enrique Ghersi (Lima, 1961) Lawyer. Professor of Law and Economics at Universidad de Lima. Director of Law Research Centre. Co-author of 'The other Path' with Hernando de Soto. Member of Mont Pelerin Society.

Jean-Louis L'hirondel

Dr L'hirondel has been a practising specialist in rheumatology at the Regional and Hospital Centre at Caen, France since 1974 and a former intern of the Hospitals of Paris. He has written widely on nitrates and their toxicity to man, co-authoring the paper, 'Man and Nitrates: the toxic myth' with his father, Dr Jean L'hirondel, was Professor of Clinical Paediatrics at Caen from 1962 to 1982. Since his father's death in 1995, Dr Jean-Louis L'hirondel has continued his professional interest in nitrates.

Christine E. Johnson

Christine Johnson is a medical editor and a freelance science journalist from Los Angeles. Her primary interest is the prognostic value of medical diagnostic tests and their use in measuring and assessing health risks. Her work has been translated into four languages and has appeared in *Christopher Street*, *Spin*, *Continuum*, and *Medicina Holistica*.

Kelvin Kemm

Dr Kemm is a nuclear physicist and gained degrees in physics and mathematics from the University of Natal. He became interested in the management of technology and in the public awareness of science. He has written many public interest articles in journals and newspapers internationally, appears often on television and radio and is a regular speaker at public events. Dr Kemm is a technology strategy consultant based in Pretoria, South Africa.

Kirill Yakovlevich Kondratyev

Professor Dr Kirill Yakovlevich Kondratyev is a Full Member of the Russian Academy of Sciences (Academician), Counsellor of the Russian Academy of Sciences in the Research Centre for Ecological Safety (St. Petersburg), and is a prominent scientist in the field of atmospheric and environmental sciences. Professor Kondratyev is the founder of several scientific schools, the author of many important books and articles. He is the co-founder and co-chairman of the Nansen International Environmental and Remote Sensing Centre Ltd (NIERSC).

James Le Fanu

Dr Le Fanu is a graduate of Cambridge University, General Practitioner in London and medical columnist of the *Sunday Telegraph* and *Daily Telegraph*. He is also a regular contributor to *The Times*, *The Catholic Herald* and *GQ Magazine*. He has written two books, 'Eat Your Heart Out: The Fallacy of the Healthy Diet', and 'Healthwise: An Intelligent Guide for the Over 60s', and three pamphlets for the Social Affairs Unit on health and equalities in national and international organisations, and environmental threats to health. In 1987, he was involved in ITV's five-part series, 'Food: Fact or Fad?' and has made several television and radio appearances discussing medical and related topics.

Michelle M. Malkin

Michelle Malkin is currently an editorial writer for the *Seattle Times*, and also serves as an occasional panellist on the PBS public affairs programme, *To the Contrary*. She was the 1995 Warren Brookes Fellow in Environmental Journalism for CEI. Before joining CEI, Malkin was an editorial writer and columnist for the *Los Angeles Daily News* where her weekly column was distributed by the New York Times News Service and published in over 35 newspapers across the country. Prior to entering print journalism, she worked as a political researcher during the 1992 presidential campaign for NBC News in Washington. In the summer of 1991, she served as a research assistant to the late Aaron Wildavsky and co-wrote a chapter on Love Canal for the book, *But is it True?* Malkin is a 1991 graduate of Oberlin College.

Lorraine Mooney

Lorraine Mooney manages health issues for the European Science and Environment Forum, and has had opinion editorials published in *The Wall Street Journal*. She has degrees in economics and medical demography.

Hector Ñaupari

(Lima, 1972) Lawyer. Professor at Universidad Nacional Mayor de San Marcos. Adviser of Peruvian Congress – Environmental Committee.

Jonathan Tolman

Jonathan Tolman is an environmental Policy Analyst at the Competitive Enterprise Institute in Washington DC. His most recent work has been on water pollution and agricultural issues. He is the author of 'Federal Agricultural Policy: A Harvest of Environmental Abuse', and 'Gaining More Ground: An Analysis of Wetlands Trends in the United States'. Prior to working with CEI, he was Associate Producer of the weekly television show, 'TechnoPolitics'. In 1991, Jonathan served as Special Assistant for the President's Council on Competitiveness, focusing on environmental and natural resource regulation. In 1992, he worked for the White House as an environmental analyst in the Office of Policy Development.

Introduction

The standard of health in the world is better than it has ever been. There is plenty of good news to report. In the industrial world almost 30 years in life expectancy has been gained in the twentieth century, almost twice the gain in all previous human history (Herman 1998). Developing countries are quickly catching up with this trend, enjoying a 60 per cent increase in life expectancy – from 40 to 64 years – between 1955 and 1995. During a similar period, child mortality, that most precise and poignant indicator of health status, fell by two-thirds in developing countries, from 287 every 1000 live births to 90. This still shows a stark contrast with the rate in developed countries of between five and six (by far the most being in the first year of life), but in actual, numbers that two-thirds equates to a decline from 21 million to 10 million children in the world dying before reaching the age of five.

At the same time nutrition has improved: 'Between 1961 and 1994, the number of daily food calories per capita increased from about 1900 to 2600 in developing countries, while their populations nearly doubled from 2.2 billion to more than 4.3 billion.' Globally in the same period, 'average daily per capita food supplies increased more than 20 per cent' (Bender and Smith 1997, Mitchell et al. 1998). Again contrary to widespread dogma, the poor benefited proportionately more than those better off. This increased production was due to technology, which bred crops to be disease resistant (giving more protection while using less pesticide), more efficient requiring less nutrient (fertiliser) per unit of output. Work is continuing by international seed companies (sometimes vilified in the West) to produce crops which require less water and repel pests (Schell 1997).

Many factors have worked together this century to produce these improvements and, while it is hard to isolate the impact of each one, they have all involved advances in science and technology. They are:

- *Clean water*
  Increased and regular food supplies and other forms of nutrition intervention, for example vitamins, food fortification, enrichment.
- *Immunisation*
  Use of pesticides and other means to control or eliminate disease vectors.
- *Antibiotics*
  Knowledge of germ theory, for example with regard to personal hygiene (DeGregori 1998).

It should be borne in mind that while these are the achievements of the advanced industrial countries, they are still the targets in many others. While the developing world is doing just that – developing – with the help of science, technology, economic growth, the West is turning its back on these things. Intellectual post-modernists enjoy the benefits, but they have forgotten what it was like before wealth, industry, chemicals and the like, improved our lives so dramatically. Chlorine and its use in disinfecting drinking water is often described as the greatest advance in public health of the twentieth century. Recently, due to an epidemiological association with certain cancers and chlorine by-products, there is a lobby to reduce or replace its use, despite there being no viable alternatives. Outbreaks of water-borne diseases have been recorded in Britain as recently as 1986, mostly caused by defective chlorination. As discovered by Peru, the risk of chlorinated water is far less than the risk of drinking untreated water (Boyce 1998).

Sufficient wealth is the best protector of life. It is always the poorest that suffer most from disease and injury. Even in England and Wales today, these effects are starkly evident. The standardised mortality ratio for cervical cancer in women in social class V (unskilled) is six times that for women in social class I (professional). For working age men in class V, the standardised mortality ratio for injuries and poisonings is nearly seven times that of men in class I. In developing countries 34 per cent of all deaths or 17 million people, die every year from communicable diseases such as acute respiratory diseases, diarrhoeal diseases, tuberculosis and malaria (WRI 1998). An expert on tuberculosis working in Birmingham has stated that he can give a reliable estimate of the rate of TB in a country by knowing its gross domestic product. Further, that no treatment will help without a secular decline in ill health from improved prosperity (Bakhshi 1998).

Cholera came to Peru for the first time in a century in 1991 and caused a devastating epidemic. The Peruvians were shocked that it could happen to them – an advanced developing nation. But the sad fact was that the authorities had not taken enough care with their water supply. There had been significant internal migration to 'young towns' around Lima, swelling the city's population by nearly a million in ten years. It was in places like these that the cholera took a grip, encouraged by lack of treated mains water, sewage and even sufficient food (see Ghersi and Ñaupari).

Neither is the risk of water-borne disease very far away in post-industrial nations and the costs of not using technology too often are not considered. An outbreak of *E. Coli* 0157:H7 occurred in one community in the United States where the spring water used in the public water system was not chlorinated (Kluger 1998). Earlier *E. Coli* 0157:H7 was spread in a swimming pool in Atlanta, Georgia that was insufficiently chlorinated. In a study in Britain covering the years 1937 to 1986, 'defective chlorination was blamed in 8 out of 10 outbreaks of disease from public water supplies and in all 13 outbreaks in private supplies' (Boyce 1998). In response to the recent outbreak of *E. Coli* 0157:H7, the community cited above, began chlorinating its water supply.

In the past few decades advanced industrial countries have been plagued by fear of technological dangers. Health scares and food scares are encouraged by the media. They are frequently illogical, hysterical and half-baked. Often a new piece of information is misunderstood or misinterpreted by a lobby group, which insists on taking an obstinately one-eyed view and demanding that something should be done. That something is usually expensive, the burden falling on somebody else like the apple industry, the beef industry or, of course, the taxpayer. Recrimination and resentment follow, but whether our health has been saved from harm by this action is doubtful, because the risks involved are minuscule (see Craven and Johnson).

Chemicals, especially if they are synthetic or 'man-made' are a common target of campaigns. A favourite is chlorine, despite the fact that 'millions of lives have been saved by the use of chlorine for disinfection of water,' there are some who would ban its further use even though the evidence for its dangers is meagre (Abelson 1994, Emsley 1994). The organochlorine pesticide, DDT is perhaps the most maligned compound. Despite its contribution to 'the greatest increase in agricultural productivity, while preventing the deaths of 500 million due to malaria' (NAS 1971) and showing only possible, not probable, adverse effects after 25 years of use by thousands of people (WHO 1979), it is seen by the public as a 'known carcinogen'.

It was its persistence in the soil that caused its demise. It was a British scientist, Dr Norman Moore, who first suggested in the 1950s that DDT was causing a decline in the eagle population. He had no part in the ensuing and continuing campaign against all chemical pesticides and does not support it. In fact, he was sprayed with DDT on being released from a prisoner of war camp in 1945, and along with thousands others, shows no ill effects. Moreover, he says, 'If I were living in a hut in Africa, I would rather have a trace of DDT in my body than risk dying of malaria' (Wakeford 1991).

His compassion is sadly not shared by others. When told that banning DDT would lead to millions of deaths, a representative of a leading environmental movement is alleged to have responded, 'So what? People are the cause of all the problems; we have too many of them; we need to get rid of them and this is as good a way as any' (Sanford 1992).

So while the poor are still struggling with the age-old, fatal problems, western societies can afford to spend time and money on environmental concerns and the identification and attempted elimination of all health hazards – both known and yet to be discovered.

Even those agencies whose job it is to improve the health of the world are torn between taking a normative stance, of setting the highest possible standards and ethics and between actively and practically relieving mortality and morbidity. An official at the World Health Organisation's policy unit expressed the following sentiments about the ethic of the new leadership under Gro Harlem Brundtland: 'Enough of economic development at the expense of mental and spiritual well-being' (Peeters 1998). Whatever this may mean, it sounds progressive and as such it appeals to people in developing countries. This is bad because, while the affluent can indulge in romantic anti-technology delusions, the developing world cannot afford to follow that lead.

The health priorities of the developed and less developed world are fundamentally different, but are being blurred by the export of ideals from western cultures. An illustration of this is given in Craven and Johnson who relate the findings of a study looking at media influence on risk perception. Two groups of people were asked to rate items on a list of some 90 hazards according to how risky they believed each hazard to be. One group consisted of Burkina Faso intellectuals, the other of French students. France and Burkina Faso are markedly different in terms of geography, economics, politics, and ethnic background, but quite similar as to media coverage. Despite radical differences in the reality of risk between the two countries, the media in Burkina Faso give coverage to hazards that may not even be present there (but would be of concern to citizens of France). The results of the study were that the Burkina Faso inhabitants had 'approximately the same preoccupations as the French respondents and to the same degree'.

This book is intended as a companion to an earlier Butterworth-Heinemann publication, *What Risk? Science Politics and Public Health*, edited by Roger Bate, which discusses the identification and effects of low-dose environmental hazards and the reaction to them. Professor Kirill Kondratyev, of the Russian Academy of Sciences, reviews these issues in this book, and appeals for a more rational approach.

We hope that this book will help to draw attention to the loss of judgement over environmental public health risks in the West, and that it might encourage developing nations to resist falling into the morass that it has created. Reactions to environmental and health scares have been expensive. In the West the costs are perhaps best measured by fractions of a percentage point of GDP in the medium- to long-term, but in developing countries they are most likely to be lives lost in the short-term.

References

Abelson, P. H. (1994). Chemicals: Perceptions versus facts, p. 183 in *Science*, **264**, No. 5156, April 8.

Bakhshi, S. (1998). 'Tuberculosis in the Caucasian population in Birmingham (1986–1996) – a disappearing disease'. Presentation given at the Sixth Annual Public Health Forum. Association of Public Health, London.

Bender, W. and Smith, M. (1997). Population, food, and nutrition, *Population Bulletin*, **51**, No. 4, February.

Boyce, N. (1998). The demon drink: Disinfecting water has saved countless lives across the world. But as recent studies show, it carries its own health risk, *New Scientist*, **157**, No. 2143, July 18.

DeGregori, T. R. (1998). 'Counter to Conventional Wisdom: In defense of

DDT and against chemophobia', Rockwell Lecture, University of Houston.
Emsley, J. (1994). *The consumer's good chemical guide: A jargon-free guide to the chemicals of everyday life*. Oxford: W. H. Freeman and Company/Spektrum.
Herman, R. (1998). Experts pondering implications of rising life expectancy, *Houston Chronicle, Washington Post* news story, May 27.
Mitchell, D. O., Ingco, M. D. et al. (1998). *The world food outlook*. Cambridge: Cambridge University Press.
Kluger, J. (1998). 'Anatomy of an outbreak: There's a deadly microbe loose on the land, and medical detectives are hot on its trail', *Time*, **152**, No. 5, August 3.
WRI (World Resources Institute) (1998). *World resources 1998–99: A guide to the global environment: People and the environment: Environmental change and human health*. New York: Oxford University Press.
NAS (National Academy of Sciences) (1971). *The life sciences: Recent progress and application to human affairs: The world of biological affairs: Requirements for the future*. Washington, DC: National Academy of Sciences, Committee on Research in Life Sciences of the Committee on Science and Public Policy.
Peeters, M. (1998). *The World Health Organisation at a Crossroads*, ESEF Publishing, Cambridge, UK.
Sanford, R. F. (1992). Environmentalism and the assault on reason, pp. 16–31 in: Lehr, Jay H. (ed.) *Rational readings on environmental concerns*. New York: Van Nostrand Reinhold.
Schell, J. (1997). A European Federation Of Biotechnology Task Group/ The Green Alliance Workshop: How Can Biotechnology Benefit The Environment? 13 January.
Wakeford, T. (1991). 'A Green in the machine: As a government scientist, Norman Moore both initiated the research that showed how pesticides damage wildlife and framed the laws that brought these chemicals under control', *New Scientist*, **132**, No. 1743, November 2.
WHO (World Health Organisation) (1979). DDT and its derivatives published under the joint sponsorship of the United Nations Environment Programme and the World Health Organisation. Environmental health criteria 9, WHO Task Group on Environmental Health Criteria for DDT and its Derivatives. Geneva: World Health Organisation.

1 Malaria and the DDT story

Kelvin Kemm

Summary

Malaria kills over two million annually, mainly African and mainly children, and causes chronic illness in survivors. Its mortality rate is growing by 5 per cent a year. Malarial areas rose by 55 per cent between 1990 and 1992. The eradication programme of the 1960s never fully tackled Africa, aiming instead for control. Now spending less than five per cent of its annual budget on malaria, the World Health Organisation has largely left the financial and technical responsibility with individual nations, often those least able to afford them. Furthermore, DDT, a cheap, reliable, safe treatment for controlling the vector of malaria – the anopheles mosquito – was banned in many countries of the world following an environmental campaign organised by the West. The rise in malaria incidence is associated with the withdrawal of DDT, yet the case for environmental damage is refuted, and the argument made for the resumption of use of DDT for malaria control.

Malaria incidence

Malaria currently affects more people in the world than any other disease, and is on the rise. In some cases, people die after a long period of suffering, but in other cases victims can be dead within a few days of diagnosis. Malaria can also affect its sufferers by recurring time and again, causing prolonged suffering over a period of years.

The World Health Organisation report of 1994 states that countries considered to be malarial numbered 90 in 1992, but had risen to 140 in 1994. The report says that 33 per cent of the world was, by 1994, classified as areas where endemic malaria has been 'considerably reduced or eliminated' but 'where the situation is now unstable or deteriorating'. The report further states that malaria affects more people in the world than any other disease – in excess of 500 million annually, with over two million deaths (WHO Report 1994).

Another startling fact is that some ninety per cent of the world's malaria occurs in Africa. As a result, we in Africa have long had the impression that the advanced first world countries are much more likely to concentrate their research and countermeasure efforts on a disease such as AIDS, which is perceived as a more immediate threat in the northern hemisphere. Indeed, the worldwide eradication programme of the 1960s, which successfully removed malaria from North America and Europe, excluded sub-Saharan Africa altogether, due to the lack of technological capability in the individual countries, and because the region's malaria problem was so huge that eradication was not considered feasible.

A change in international policy from eradication to control of malaria, in which quick, dramatic, successful results were replaced by long-term, careful management showing limited success, meant that international funding for malaria activity, which had amounted to US$100 000 million in 1960, was largely withdrawn by 1974. The WHO, for example, which had committed 30 per cent of its annual budget to malaria during the eradication years, currently allocates less than 5 per cent. This left the financial and technical responsibilities for malaria control almost entirely with the individual nations, many of which did not have adequate resources, technical expertise, or the administrative infrastructure to carry out effectual programmes. The implication is that Africa will have to champion the fight against malaria, and not rely on first world initiative and leadership.

In South Africa, malaria cases as reported by the South African Department of Health, were 1623 in 1974, with 16 deaths (the year DDT was banned for agricultural use) but 20 years later, by 1994 had risen to 13 000 (SA Dept of Health, 1994). In 1996, total officially recorded malaria cases were 27 035 with 163 deaths. In 1997, the picture improved slightly, to 23 116 cases, and 104 deaths. In the first half of 1998, the incidence of the disease progressed worse than 1996 and 1997, with half-year total deaths being 111 (SA Dept of Health 1998).

Control of malaria

The first and most basic priority in malaria control is to prevent infected individuals from becoming severely ill and dying. The fatality rate for cerebral malaria in children can be 50 per cent, so the urgency is clear. Then, measures should be taken which limit the human–mosquito contact for people living in endemic areas, such as bed-nets, screens, repellents, and coils which burn overnight, releasing insecticide, when the mosquitoes feed. The best final outcome is an anti-malarial vaccine, however, the most practical long-term ideal is to control or eradicate the vector of the disease, the anopheles mosquito itself, which nurtures and then transfers the malaria parasite to its victim.

Anti-vector measures are the most effective tools for preventing and controlling malaria transmission. Common measures include indoor and outdoor insecticide spraying; the use of insecticide-impregnated bed-nets and curtains; the killing of mosquito larvae in pools of standing water, flushing the pools, or draining them, as practised since the time of the Greeks and Romans; and designing and locating houses and animal shelters in areas with the least possible exposure to malarial transmission. Application of these measures can, alone or in combination, reduce vector abundance, human-vector contact and vector infectivity, but the greatest impact is seen with the reduction in vector infectivity. In view of its ability to achieve these vector controls, insecticide spraying aimed at killing adult mosquitoes is the anti-vector approach strategy of choice in malaria control, but it cannot be used indiscriminately. Spraying insecticides with a residual effect inside houses and cattle sheds creates an inhospitable environment to both the mosquitoes and the malaria parasites developing within them. The use of insecticides to directly control or, if possible, to eliminate the mosquito vectors is currently the most practical and cost-effective tool. Sometimes it is the only practical way to overcome them.

Malaria vaccine

Current work being carried out by Professor Manuel Patarroyo, Director of the Institute of Immunology, University of Colombia, Bogota, on a new range of malaria vaccines, is encouraging, and should undoubtedly be supported by the international community, but this does not address the malaria vector, but rather aims at the patient or potential patient. Professor Patarroyo has produced the first chemical vaccine, and also the first vaccine for malaria

(Spurgeon 1995). Colombia, like areas of Africa, also has a significant malaria problem, and so there is a domestic need to aggressively attack the disease.

Mosquito insecticide

In the past, the most effective anti-mosquito pesticide was DDT. It was effective, easy to use, and inexpensive. But it was banned. This action was primarily due to political pressure from some environmental organisations, notably in the USA. Unfortunately, much of the anti-DDT propaganda and claims concerning its supposed dangers were inaccurate and premature, and sometimes appeared to have been contrived to gain political advantage.

The resulting ban of DDT in the USA and other first world countries resulted in the spread of an essentially forced ban of the pesticide to many other countries, including South Africa. Since the banning of DDT, malaria has increased rapidly, as have malaria-related deaths. In the past year in southern Africa, the increase in malaria cases has been alarming. They are not only increasing in total numbers but are also being found in cities such as Pretoria and Pietersburg in South Africa, and Bulawayo in Zimbabwe, which are far away from the actual malarial areas. This phenomenon is attributed in part to infected persons moving into the cities, and so leading to the infections of others. In northern Namibia, malaria cases in the first half of 1996 compared to 1995 increased by 40 per cent and children under five years of age made up 30 per cent of the deaths. In Zimbabwe, malaria is increasing dramatically, and in 1997 there were approximately 1.2 million cases with 2500 deaths. Accurate figures for Mozambique and Angola are not known. The threat covers a wide region (Gittens 1995).

Is the DDT ban justifiable?

Recently in Zimbabwe a Health Ministry spokesman, Dr Paulinus Sikhosana, was quoted as saying that Zimbabwe intended to reintroduce DDT in an attempt to reduce the spread of malaria. He said that they intended to use 10 000 tons of DDT. Dr Sikhosana also said that he did not quite know how the DDT ban would be overturned. He implied that Zimbabwe feared international pressure. This announcement resulted in predictable anti-DDT campaigning in Zimbabwe and in South Africa.

DDT history

DDT is a powerful, effective insecticide that was originally discovered chemically in 1874, but its insecticide properties were only discovered in 1939 by Paul Müller, a chemistry researcher working in Switzerland. As the Second World War intensified Müller revealed his discovery to the British, who classified it as a war secret.

The first patents were taken out in Switzerland and Britain in 1942. DDT immediately changed the face of the world, even though it was a closely guarded secret, for during the First World War more soldiers had died from typhus than from enemy bullets. DDT eliminated this problem in the Second World War, by killing body lice. When Nazi concentration camps housing Jewish prisoners were liberated after World War II, all the prisoners were found to be infected with lice. DDT was used on all prisoners and within days the lice infestation was wiped out.

In 1948 Paul Müller was awarded the Nobel Prize for the discovery and development of DDT, which was hailed as the miracle pesticide since it was so selective in its action. DDT had replaced earlier pesticides, such as arsenic and nicotine, that were toxic to all life. DDT however, was highly effective against target pests, such as lice and mosquitoes, but had no effect on humans, mammals, or indeed other larger insects, earthworms, and other soil organisms. Because it was so target specific, DDT was also used to improve crop yields dramatically by killing off agricultural insect pests.

As an example of DDT's effectiveness, annual cases of malaria in Sri Lanka in 1948, were 2.8 million. Then DDT was introduced to suppress the malaria-carrying mosquito, and by 1963 the annual number of cases had fallen to 17. Worldwide, the introduction of DDT had a dramatic effect on reducing malaria. For example, in Natal (a warm coastal province in South Africa) in 1932, the estimated annual death toll from malaria was between 10 000 and 22 000 (Smit et al. 1992). The later introduction of DDT eradicated this menace.

The environmental backlash

Then an attack was launched on DDT by extremist 'greens' in the USA. This was championed by the then relatively insignificant US Environmental Protection Agency (EPA). The success of the anti-DDT campaign has thrust the EPA into the position of political power which it enjoys today.

The campaign was so successful that DDT is essentially banned in large parts of the World except for certain special uses. There has been dramatic increase in pests directly because of the ban. In Sri Lanka, due to worldwide reaction, DDT spraying was stopped in 1964, and by 1969 malaria cases had increased from the 17 cases to an annual figure of two and a half million. Professor Kenneth Mellanby, first Director of the Monks Wood Experimental Station in the UK, made the observation that in Sri Lanka DDT caused no harm to the spraying workers, but when DDT was replaced by parathion which was preferred because it was not persistent, even though it is highly toxic, many deaths amongst spraying workers occurred.

Environmental Protection Agency

The EPA indictment against DDT was based on three premises:

1. DDT was a danger to birdlife, particularly raptors.
2. DDT was carcinogenic in humans.
3. DDT had an indefinite lifetime in soil and so would pose a permanent threat to insects and various other organisms.

However, since these charges were made, science has been able to research them more thoroughly, and current thinking has changed. Responses to the above points are as follows:

1. Throughout the period of use of DDT in the USA, the number of raptors continued to increase. An official count in Pennsylvania of ospreys recorded 191 in 1946 and over 600 by the time DDT was banned.

Great media coverage was given to the vociferous green activist, Rachel Carson, who claimed that the American robin was being virtually exterminated by DDT. Carson's book *Silent Spring* popularised the anti-DDT scare (Carson 1962). In fact, it has since been established that the overall robin population of the US actually increased during the period of DDT usage. The US Audubon Society, dedicated to the protection of birdlife, recorded an increase in robin population of 12 per cent between 1941 and 1972.

There were incidents in which DDT-induced bird egg shell thinning damaged the breeding cycles of birds. However, close examination revealed that these incidents were limited to small, localised areas in which DDT had been used in an irresponsible manner, far in excess of recommended doses.

Egg shell thinning had been observed for decades before DDT was used (Milius 1998), and many causes are known, such as diets low in calcium or vitamin D, high nocturnal temperatures and others. Moreover, experiments designed to test toxic effects of DDT failed to find a result supporting the hypothesis of harm, even though birds were fed 6000 to 20 000 times more DDT than the typical 0.3 parts per million normally found in their food.

Quail that were fed 200 parts per million in all their food throughout the entire breeding cycle nevertheless hatched 80 per cent of their chicks in comparison to 84 per cent in the control group. No egg shell thinning was reported. In the case of pheasants treated the same way, there seemed to be a benefit. The DDT-treated birds hatched 80.6 per cent of their chicks, while only 57.4 per cent hatched in the control groups.

In South Africa, no DDT-induced interruption of bird breeding cycles has been detected, and no bird egg shell thinning due to DDT is known; as reported by the Head of the Department of Birds at the Transvaal Museum (Kemp 1995).

2. In 1978, the US National Cancer Institute declared that DDT was not a carcinogen. No human carcinogenic effect

has ever been detected. In fact, the dose needed for humans to be poisoned (about 100 grams) is so high that almost all poisonings of humans by DDT have been accidental or suicidal (Nhachu and Kasilo 1990).

No systemic poisoning of humans has resulted from occupational exposure to DDT (Violante and Coltelli 1986, Brown and Chow 1975, Saxena et al. 1987, Nhachi and Loewerson 1989).

In South Africa, although DDT has been detected in minute amounts in mother's milk, no negative effects were detected (van Dyk et al. 1987).

3. DDT does not have an indefinite life in soil. It is broken down to DDE, and this continues to break down further to less complex molecules. Obviously, the persistence of DDT in soil (from about six months to about three years) was hailed as one of its major positive points from an insecticide point of view. However, it certainly is not indefinite, otherwise no respraying would ever be necessary.

   DDT is also broken down in the atmosphere and in water by mechanisms not yet fully understood. DDT is extremely insoluble in water and so the potential for movement in groundwater and other water-borne methods, is low.

Human diseases

The vast and widespread use of DDT on humans is evidence of its positive effects and lack of claimed negative effects both short term and long term.

There are many powerful illustrations of malaria reduction. In Italy, in 1945, a five-year plan to eradicate malaria was started. Spraying with DDT started on 5 March 1946 and continued until mid-May. The country was divided into zones. Figures from capturing stations set up to monitor mosquitoes are dramatic.

In June 1945, maximum figures of 50 000 mosquitoes per station were recorded. In June 1946, the figures were 200 and in 1947 and thereafter most stations recorded zero. Writing in 1957, S. W.

Simmons stated that not a single death occurred from malaria in Italy after 1948. Similar success stories were recorded all over the world, such as the already-mentioned case of Sri Lanka.

After the War, the use of DDT against human body lice continued with startling success. Typhus continued to be epidemic in Egypt, so DDT was extensively used after the war and by 1948 the incidence was reduced to only 187, from a maximum of 40 188 in 1943.

A similar situation existed in post-war Tokyo. Between 16 and 19 February 1946, a total of 1 837 511 people were treated with DDT and a further 1 306 360 were dusted during the next 14 days. A total of 91 708 kg of DDT was used. Body lice were wiped out.

In 1951, there was a nation-wide campaign in Mexico to eradicate the body louse. A dust containing 10 per cent DDT was widely used. It was also found that soap containing 3.2 per cent DDT was effective in killing lice. In the years after 1951, no further cases of typhus were reported.

Very high louse infestations (89 per cent of those examined) were found in Colombia. Dusting with DDT at four monthly intervals reduced this level to 2 per cent. When the dusting was stopped, within 10 months infestation was back up to 60 per cent.

Therapy

It is interesting to note that various therapeutic uses for DDT have been found. A 17-year-old patient had suffered from a genetically-determined jaundice from his thirteenth year. This produced a high increase in the bilirubin content of the blood, resulting in nausea.

He was given a daily dose of 90 mg of DDT for six months. His high bilirubin count fell to an acceptable level immediately, and stayed low for at least 7 months after the DDT treatment was discontinued. This is attributed to low continuous release from the patient's body fat. No adverse side effects were detected.

DDT has also been found valuable in treating the rare disease of inoperable adrenal cortical carcinoma (generic name of 'mitotane'). It has been found to prolong patients' lives by 7 to 8 months on average. Very high doses of between 7 to 285 mg per kilogram of body weight per day are used. The average dose is approximately 100 mg per kg per day. The highest doses caused nausea, vomiting, and apathy, but no injuries to the liver, kidneys or bone marrow.

Human exposure

Perhaps the most convincing fact regarding high exposure to DDT is that since 1942, millions of people have been exposed to DDT for various public health purposes. There has not been one single death and no convincing evidence of any serious human illness (Spindler 1983).

DDT has been found in many foodstuffs in all countries where it has been used. This provided considerable ammunition to the anti-DDT lobby who claimed that poison was everywhere. In fact, in the USA the highest level of DDT found in a 'market basket survey' was only about a quarter of a milligram per day, a level far below that which might be toxic, and lower than most of the ultra-conservative recommendations for a 'safe' level (Spindler 1983).

Furthermore, there has been no detected increase in cancer in any group involved in malaria control programmes, in agriculture, or in the chemical industry. In 1967, E. R. Laws reported his findings on 35 workers in the Montrose Chemical Corporation's plant in the USA. They had worked for between 11 to 19 years in a plant producing nearly 3000 tons of DDT per month. The workers had suffered an exposure to DDT estimated at 450 times greater than that of the average citizen outside the plant. Their medical histories, x-rays, and clinical laboratory tests revealed no ill effect which could be attributed to exposure to DDT.

In 1979 an expert group of the World Health Organisation said: 'more than 30 years after the introduction of DDT, there is no evidence, whatsoever, that DDT is carcinogenic in man' (WHO 1979).

Eco-militancy and the law

In 1967 a new organisation, the Environmental Defence Fund (EDF) was formed 'to preserve the environment through legal action backed by scientific testimony' (EDF 1967). It was led by a flamboyant lawyer Victor Yannacone who gained international attention for his skilful cross examination of witnesses and by his powerful advocacy of his case.

The formation of the EDF, with its tactics of confrontation, was a watershed in the opposition to DDT. The EDF abandoned the genteel tradition of the conservative approach to science that had previously been used by important bodies such as the Audubon Society.

Yannacone was extremely aggressive, using an approach that would have been considered improper in the UK or South Africa where barratry (that is, incitement to vexatious litigation) can result in disbarment from the profession. But in the American legal climate, this approach served as a rallying cry for those who wished to join his cause.

Yannacone explained that their goal was not to ban DDT in any particular place but to find a public forum with an independent arbiter before whom to get a judgement condemning DDT as an environmental pollutant. The EDF also wanted maximum publicity. A group in Wisconsin, the CNRA (Citizens' National Resources Association) applied for a 'declaratory ruling' in which the relevant state agency was required to hold public hearings in which each side would present its case, subject to the rules of evidence and cross examination. The CNRA had worked closely with the EDF.

On 28 October 1968 the CNRA asked the Department of Natural Resources for a declaratory ruling on DDT. The chief hearing examiner of the department, Maurice van Susteren, assigned himself to conduct the hearing. The hearing was expected to last 10 days, but in fact took much longer and went on until May 1969.

The EDF were well prepared but soon ran out of money. However, public enthusiasm ensured that the required cash was soon raised. On the other hand, the chemical industry was ill prepared and made a bad showing, even when the facts supported their case. During the hearing the Wisconsin legislature banned the use of DDT in the state.

For some unexplained reason, van Susteren's findings were not filed for a year after the hearings. Van Susteren found that DDT was mobile, stored in fat, and was present in all levels in the food chain. He said that the bioconcentration of DDT made it impossible to establish safe levels of exposure or safe methods of use. He ruled that 'the only valid permissible inference is that DDT in small doses has a harmful effect on the mammalian nervous system'. He said that it was clear that DDT was having an effect on the environment – it was thinning eggshells in certain susceptible species.

While acknowledging that DDT had provided enormous economic benefits, the examiner ruled that this was not a topic for this investigation.

Finally, he ruled that DDT was a pollutant within the meaning of the Wisconsin statute. This finding caused DDT to be banned throughout most of the USA. Although most of the findings were directly contradicted by the scientific evidence, the findings had been arrived at by rules of court procedure, not by rules of scientific procedure.

The chemical industry was horrified and promptly, under the Federal Insecticide Fungicide and Rodenticide Act (FIFRA), protested against the court findings. This resulted in the Consolidated DDT Hearing which began in August 1971 and continued until the following March. Like the previous hearing, it was conducted like a trial. The hearing examiner was Edmund Sweeney. His most notorious action was to demand that a scientist give a 'yes' or 'no' answer to a question which could not be answered in this way. As a result, a number of government scientists were banned from giving evidence at the hearings. But the issue was resolved eventually. This time the chemical industry was much better prepared. In the end Sweeney essentially reversed the findings of the previous hearing and came to the following conclusions:

1. DDT is extremely low in acute toxicity to man.
2. DDT is not a safety hazard to man when used as directed.
3. DDT can have a deleterious effect on freshwater fish and estuarine organisms when directly applied to the water.
4. DDT can have an adverse effect on beneficial animals.
5. DDT is concentrated in organisms and can be transferred through food chains.
6. DDT is not a carcinogenic hazard to man.

7. DDT is not a mutagenic or teratogenic hazard to man.
8. DDT used under the registrations involved here does not have a deleterious effect on freshwater fish, estuarine organisms, wild birds, or other wildlife.
9. There is a present need for the continued use of DDT for the essential uses defined in this case (Sweeney 1971).

This was considered a reasonable finding and should have allowed DDT to be used, with stringent safeguards, in the USA. But this was not the end of the matter.

Two months after Sweeney had published his findings, William Ruckelshaus, the administrator of the Environmental Protection Agency (EPA) banned all remaining uses of DDT with the exception of those deemed essential for specific public health purposes. Ruckelshaus had not attended any of the hearings. Nevertheless, he completely overturned Sweeney's findings and ruled:

1 DDT is lethal to many beneficial insects.
2. DDT can have lethal and sub-lethal effects on useful aquatic freshwater invertebrates.
3. DDT is toxic to fish.
4. DDT can affect the reproductive success of fish.
5. DDT can cause thinning of bird eggshells and thus impair reproductive success.
6. DDT is a potent human carcinogen.
7. Responsible scientists believe that tumour induction in mice is a valid warning of possible carcinogenic properties.
8. There are no adequate human epidemiological data on the carcinogenicity of DDT, nor is it likely that such can be obtained.
9. DDT presents a carcinogenic risk (Ruckelshaus 1972).

It is interesting to note that two years earlier, the same William Ruckelshaus had made a completely different statement and said:

> DDT is not endangering the public health and has an amazing and exemplary record of safe use. DDT, when properly used at recommended concentrations, does not cause a toxic response in man or other mammals and is not harmful. The carcinogenic claims regarding DDT are unproved speculation (Ruckelshaus 1970).

It is difficult to imagine how a senior administrator could come to such conclusions. At the time, Ruckelshaus denied that the decision to ban DDT was not scientifically based, but at a later date he said:

> Decisions by government involving the use of toxic substances are political with a small 'p' – science, along with other disciplines such as economics, has a role to play. The ultimate judgement remains political. In the case of pesticides in our country, the power to make this judgement has been delegated to the administrator of EPA (Ruckelshaus 1972).

The current situation

The attack on DDT in the USA had the effect that the USA proceeded to pressure other countries to follow their lead. This was done by various methods, one of which was threatening not to import agricultural produce and related goods if DDT had been used.

As a result, countries such as South Africa introduced bans even though no local scientific decision was taken in this regard. Decisions taken were economic and political. In South Africa, DDT was banned for agricultural use in 1974 but farmers were allowed to use stockpiled supplies until 1976. For malaria control, DDT may only be currently used in South Africa in limited quantities inside informal dwellings, and only under strict government control. The porous walls of these dwellings are unsuitable surfaces for effective DDT action, since DDT is a contact poison, and most of it seeps into the walls. Spraying inside informal dwellings is carried out infrequently and on a small scale.

Zimbabwe has been the first southern African country to tentatively voice an intention to reintroduce DDT only to be met by protests from radical interest groups.

With the relaxation of political tensions in southern Africa, more people are crossing from Angola and Mozambique to and from South Africa. The same is true of the borders between other states. The potential for malaria spread is therefore greater than ever before. No doubt the same is true of human body lice and other pests, against which DDT is effective. Maybe it is time for the developing world to not succumb to politically induced decisions of the first world, particularly those that affect our livelihood and not theirs.

References

Brown, J. R., Chow, K. Y. (1975). *Bull Environ Contam Toxicol* **13**: 483.

Carson, R. (1962). *Silent Spring*. New York: Houghton Mifflin.

Environmental Defense Fund (EDF) (1967).

Gittens, C. (1995). *Farmers Weekly*: **17** November, 16.

Kemp, A. (1995). Transvaal Museum: Balancing Bodies, Biocides and Birds (unpublished).

Laws, E. R. (1967). Montrose Chemical Corporation.

Milius, S. (1998). 'Birds' eggshells started to thin long before DDT', *New Scientist*, **153**, 261.

Nhachi, C. F. B., Loewerson, R. (1989). *Bull Environ Contam Toxicol* **43**: 493.

Nhachi, C. F. B., Kasilo, O. J. (1990). *Bull Environ Contam Toxicol* **45**: 189.

Ruckelshaus, W. (1972). 'Consolidated DDT Hearings: Opinion and Order of the Administrator,' *Federal Register*, **37**, July 7.

Saxena, S. P., Khare, C., Farrog, A. et al. (1987). *Bull Environ Contam Toxicol* **38**: 392.

Simmons, S. W. (1957). Malaria in Post-War Italy.

Smit, H. E., Bouwman, H. and Le Sueur, D. (1992). *DDT and Community Health* **3**:3.

South African Department of Health Bulletin (1994). Nov.

South African Department of Health: Private Communication (1998). April.

Spindler, M. (1983). *Residue Reviews*, **90**: 1–34.

Spurgeon, D. (1995). *Southern Lights: Celebrating the scientific achievements of the developing world*. IDRC Books June.

Sweeney, E. M. (1972). Hearing Examiner's Recommended Findings, Conclusions, and Orders, consolidated DDT hearings, (Washington, DC: Environmental Protection Agency, Apr. 25), p.93.

Van Dyk, L. P., Lotter, L. H., Mullen, J. E. C. and de Kock, A. (1987). *Chemosphere* **16**, 4:705.
Violante, F. S. and Coltelli, F. (1986). *Arch Environ Health* **41**: 117.
World Health Organisation (1994). *World Health Report* CTD/MIP/ 94.4:5.

2 Dirty water: Cholera in Peru

Causes and those responsible

Enrique Ghersi and Hector Ñaupari

Summary

In the late 1980s the water supply system in Peru was old or inadequate and the costs of repair and investment were too much for the Government to bear. Peru insisted that water was a public good and retained control. As a result of poor management and lack of responsibility, large sectors of the population were without clean water, and were malnourished. Distribution via water trucks became corrupted, with water authorities and water truck suppliers charging illegally for water. Chlorination of water was below the WHO recommended levels in most areas, partly because of fears about carcinogenic effects of chlorine by-products raised by the US Environmental Protection Agency. The almost inevitable cholera outbreak lasted from 1991 to 1996 killing more than 6000 and afflicting over 800 000. The epidemic was aggravated by lack of governmental leadership, or sufficient money to run the medical and educational campaign planned, and by the general poor state of health and hygiene among the poorer workers.

Introduction

The 1991 cholera epidemic, which cannot yet be considered as ended, is the largest epidemic of any disease in the twentieth century in Peru (Wachsmuth et al. 1991). Tracked by the Pan-American Health Organization, cholera rapidly crossed Peru's borders and, by the middle of the year, had reached Ecuador, Colombia, Brazil, Chile and Guatemala, and seriously threatened other countries of the region.

The epidemic put the Americas on the list of regions to which the seventh cholera pandemic has spread. This pandemic, which originated in Indonesia in 1961, has swept across the world, and the World Health Organisation believes that it has affected no fewer than 98 countries.

The initial hypothesis prompting this research was that among the most important causes of the cholera epidemic was the substantial reduction in the chlorination of water for human consumption by governmental water authority, SEDAPAL and the health authority DIGESA, due to an erroneous water management policy, based on state interventionism and influenced by borrowed environmental concerns.

The factors which decisively conspired against controlling the epidemic and even encouraged its propagation at unexpected levels were the complete absence of maintenance of the system of water for human consumption, the lack of cleanliness in water wells and domestic containers, the lack of access on the part of a substantial portion of the population (47 per cent) to properly chlorinated potable water, the discontinuity of the water chlorination process in the system's end points, due to deficient installation of the service.

In the face of the severity of these problems, none of the authorities responsible for the management of water for human consumption in Peru had any certainty of the quality of the water sent to the population. A level of chlorination from 0.2 to 0.3 mg/l of chlorine was added to the water, far below the level utilised internationally. There was no analysis of the efficacy of this level, which, as was discovered sometime later, was inadequate to preserve the health of Peruvians during the epidemic.

That is to say, if the level of chlorination had been kept at the level of 0.8 mg/l recommended by the WHO, the incidence of the epidemic wouldn't have been so explosive (Haratani and Hernández 1991).

No cases of cholera had been reported in Peru for a century. Notwithstanding this, the serious deterioration in the water treatment system could not have gone so blatantly unnoticed. A large part of the responsibility fell upon the public officials in charge of the treatment company. Nevertheless, the responsibility of ecological groups who insisted on campaigns for the protection of endangered animals or groups highlighting the plight of native rural populations, instead of alerting the population to this serious problem, should also be mentioned. These groups called for the disinfection of water by chlorine to be discontinued, and succeeded in influencing the water authorities to reduce amounts used. An attempt was made to repair this negligence, but it was obvious that the priorities of these organisations ignored the health of the population.

Cholera in Peru

At the end of January 1991, the summer in the Southern Hemisphere, the cholera epidemic finally arrived in Latin America. The outbreak, which is believed to be part of the Seventh Great Pandemic, began on the coast of Peru (Wachsmuth et al. 1991).

The cholera suffered by Peru that year became the biggest epidemic in our country in the twentieth century due to the unexpected virulence and the velocity of spreading. In terms of human lives alone, the cholera epidemic left Peru with more than 6000 dead and 800 000 cases (Suematsu 1996, Graham and Gray 1997).

From time immemorial, cholera, along with the plague, small pox and yellow fever, has been one of the four scourges responsible for the largest disease epidemics throughout the history of man. This is a disease caused by the bacterium *vibrio cholerae,* which settles and reproduces in the small intestine (Elmore Luján 1991). By means of the production of a potent toxin, the bacterium causes an active secretion of water and electrolytes in the intestinal tract, of which the clinical symptoms are diarrhoea and vomiting. In contrast to other micro-organisms which also cause diarrhoea, cholera vibrio always produces a more severe and abrupt illness. An adult affected by this disease can defecate up to two litres per hour, which, if not replaced rapidly, may cause death by dehydration.

The diagnosis of typical cases does not cause difficulties, but one cannot forget that cholera can simulate any type of intestinal infection, serious or not. The 'iceberg phenomenon' is also characteristic of cholera, latent infections are much more numerous than manifest cases.

According to the Pan-American Health Organization, in its publication, 'The Control of Transmissible Diseases in Man', cholera is defined as an 'acute bacterial intestinal disease which is characterized by its rapid beginning, watery and profuse diarrhoea, occasional vomiting, rapid dehydration, acidosis and circulatory collapse. The asymptomatic infection is much more frequent than the appearance of the clinical tableau, especially in the case of El Tor biotype organisms. Untreated slight cases, in which there is only diarrhoea, are common, especially among children. In untreated serious cases, the subject may die within a matter of hours, and the mortality rate exceeds 50 per cent; with the appropriate treatment, the rate is less than 1 per cent (PAHO 1987).

Cholera can be classified as an environmental disease, given that the bacterium propagates fundamentally by the ingestion of water and food products contaminated with faeces or the vomit of patients or, to a lesser degree, the faeces of carriers. The bacterium can propagate rapidly in regions that lack sewage networks, clean potable water or adequate sanitary services, and directly contaminates drinking and irrigation water. Food is contaminated by dirty hands, and flies play a very important role in the transmission of the disease. An example of this is the ingestion of raw or poorly cooked shellfish and fish coming from polluted waters, which caused epidemics in Guam and Portugal, and sporadic cases in the United States.

An essential factor for the spreading of cholera is the poor condition of environmental health, especially the lack of sufficient drinking water and the lack of sanitary facilities for the adequate disposal of excrement.

The first time in the history of epidemiology when a source of drinking water was related directly as a cause of a cholera epidemic, was the research performed by Dr John Snow in the so-called Broad Street epidemic in London, in 1854 (PAHO 1991). He found that drinking water had been contaminated with sewage waters which contained discharges from people suffering from cholera. At the end of the investigation, Dr Snow concluded that intestinal discharges from cholera patients carried something that had contaminated the water and had become the direct cause of the epidemic.

The overwhelming proof that the best defence against the disease of cholera is having a good water supply with a water filtration process in its facilities, using sand beds, was found during the Hamburg cholera epidemic in 1892. At that time, Hamburg was supplied with water from the Elbe River, but its supply system facilities did not have any filtration process. On the other hand, the city of Altona, practically an extension of Hamburg, was also supplied with water from the Elbe River, but its supply system did have a filtration process. Hamburg was swept by cholera, while Altona did not record a single case.

The most important reservoir for this disease has been in Asia. The Ganges and Brahmaputra deltas and the Bengal region in general, have been distinguished as natural centres for the disease. It is from there that the six pandemics of classical cholera have spread across the world.

The sixth pandemic saw its decline from 1923. An outbreak was discovered in the 1930s El Tor biotype, on the island of Sulawesi. In 1961, this locus exploded and began to expand rapidly, beginning the seventh known pandemic of cholera. This extended to the Philippines, then north toward Taiwan and Korea, and to the west, to India, Pakistan, the Middle East and Europe. It also arrived in eastern Africa and the islands of the Pacific (PAHO Ecuador 1991). The seventh great pandemic, as we have seen, began in Peru.

Water supply and sanitation

In 1980, Lima's population was estimated at 5 020 000 persons, of whom 3 778 000 received water from the mains system and 1 242 000 from vendors. It was estimated that there were 500 000 unauthorised connections, with an average of 7.5 persons per connection (Haratani and Hernández 1991).

By 1991, Lima's population had mushroomed to an estimated 6 000 000, of whom approximately 4 700 000 received water from the mains system and 1 500 000 from water vendors. It is estimated that there were 665 000 unauthorised connections.

Around 1990 in Lima, between 20 per cent and 25 per cent of the population did not have running water, and were supplied mainly through water trucks; before the beginning of cholera, there were 13 water supply trucks. They belonged to SEDAPAL, although they were managed by several municipal councils, which sold water to the truckers at a higher price than potable mains water. At the same time, the owners of the distribution trucks raised the prices, taking advantage of the scarcity.

It should be added that, before the epidemic, there was no quality control over the water distributed by means of these trucks.

Indeed, in 1991 the water and treatment sector in Peru was in a mess, facing enormous delays in the achievement of its goals. What was, and is, serious is that disease originates within this system, due to its poor service conditions, thus creating deficient health situations that are very dangerous to the Peruvian people.

With respect to the lack of sanitary services that same year, the figures speak for themselves: at the national level, only 35 per cent of the population had toilets and in rural areas this figure was 2 per cent. Approximately 25 per cent of the rural population had rudimentary water-piping systems. None of these had sewage systems. Even when some families had installed cesspool latrines, more than three-quarters of the rural population did not have formal excrement-evacuation facilities. In the same manner, at the national level, 42 per cent of the population disposed of their trash in a river or field, and in rural areas this percentage rose to 78 per cent. The most serious part is that this situation has not changed substantially today (Tejada de Rivero 1991).

The daily production of water was 1 229 000 cubic metres. Of this water production, 70 per cent comes from surface waters and 30 per cent from deep wells. Although this is the water production, there are no valid statistics for water consumption. It is estimated that the system had losses of at least 20 per cent, and maybe as much as 40 per cent of water production.

The only surface source of water is the Rímac River, which originates in the highlands of the Andes and flows 125 kilometres to the sea in Lima, with a fall of 5000 metres. Although it originates as a good source, mining operations, extensive landslides and erosion, direct discharge of untreated residual waters and industrial waste along many points, rapidly change it into a highly polluted water source which is difficult to treat. The water is drawn at the La Atarjea plant within the limits of the City of Lima. La Atarjea is a complete treatment plant, designed to treat 15 cubic metres per second. Due to the nature of the difficulties in treating the water source, this plant is limited to 9 cubic metres per second, that is, 60 per cent of its real capacity. This is worrisome, inasmuch as the plant provides 60 per cent of the water for Lima.

Pre-treatment consists of an intake with grate, primary sedimentation and pre-chlorination before the water passes through a large storage reservoir with a capacity of 50 000 cubic metres. Then the reservoir's water goes to primary and secondary sedimentation facilities, is filtered through sand, is chlorinated and proceeds to the storage reservoirs before entering the distribution system. The operation of this treatment facility is extremely difficult.

The central system has large feeder conduits which go to the different areas, which are then divided into service pipes for the individual separate areas. As has already been stated, general water losses are between 20 per cent and 40 per cent. The majority of these losses, are probably due to broken or leaky faucets and valves and primary conduits in the city portion, which are between 40 and 60 years old, and are deficient. As one drives through the city, it is common to see flooded areas which appear to be caused by breaks in the conduits. It seems that little attention is paid to repairing them.

Of the 765 000 total existing connections, only 400 000 actually receive regular daily service. The others are subject to frequent shortages and cut-offs. There is little effort or incentive to conserve water and there are no direct efforts or inspections which assure the absence of interconnections in home, business and commercial facilities.

Due to the frequent water shortages, almost all water services have some type of storage facility, ranging from simple and poorly-constructed to large and well built. They are open and unprotected and are subject to local contamination. Tanker services are common and are likely to cause problems. Public and private tankers fill their water tanks in designated storage points along the system, and then they transport them to outlying areas. There is usually home service in these areas, where each person has to purchase water. The tanker service water can be a danger for the population served, due to the possible contamination of the tank's content or the home owner's container.

In the south of Lima, there is extensive use of untreated residual water and effluents from lagoons for different types of irrigation, which includes use in gardens. It is not known if these residual waters are used in the home.

In the young towns of the coast, today as in the epidemic period, the people supply themselves with water through distribution trucks, and their waste is discharged by burying it, wherever and whenever possible.

In the urban areas of large Peruvian cities, the sewer systems have a series of deficiencies, and the discharge from the collectors is very often used as a source of irrigation water without treatment and practically without any control. As a result, there is a high level of contamination of agricultural food products, especially vegetables.

In the marginal areas of the cities and in rural areas, the inhabitants without access to excrement removal system defecate outdoors, and their faeces, in one way or another, become food contamination points.

It is not surprising, then, that a special report prepared by the United States Agency for International Development (USAID) Mission in Peru in May 1991, (Water and Sanitation for Health (WASH) Haratani and Hernández 1991) prompted by the cholera epidemic in Peru, in which the infrastructure, water supply, excrement removal and sewage in Peru was examined and evaluated, summarised and confirmed this situation with the following frank and true words:

> The conditions of the water supply and treatment in Peru have been described and characterised as a disaster waiting to happen at any time. The cholera epidemic which Peru and other countries are suffering only put a name on the disaster. In a certain sense, nature, according to its own evolutionary sense, cold and deliberate, is currently collecting its part of a debt which the world has not paid, a debt in the form of inadequate investments in the water and treatment infrastructures and in health education, which is an inseparable part of them.
>
> (Haratani and Hernández 1991)

This biting but just observation describes the condition of the water supply and excrement disposal in the country's urban and rural areas. Responsibility for these programmes falls on the Ministry of Health and the Ministry of Housing and Construction.

Before the beginning of the cholera outbreak, there was regular chlorination only in Cajamarca. In all the other cities, either there was no disinfection or it was sporadic and performed without quality control, with inadequate equipment and shortages of money and supplies. Furthermore, the deterioration of the public system, low pressure and continuous changes in pressure in the distribution system, the many covert connections and the inadequate water storage practices in homes, exposed the water to a long-term risk of contamination.

The lowest level of chlorination prior to the outbreak of cholera occurred in the city of Iquitos, with 0.1 mg/l. The condition of the drinking water treatment and distribution system was very similar to that of Lima: high contamination of the source water; a difficult-to-manage and poor-quality system, lack of process control; irregular application of chlorine; a network in poor condition, comprising breaks, leaks and the combination of treated and used waters; the lack of quality control prior to the outbreak; the creation of covert connections, the dumping of drainage without any type of treatment.

The most important part of the USAID report about drinking water in Lima and other coastal cities was that, prior to the outbreak of cholera, there was no regular monitoring of the system for residues of chlorine for bacteriological purposes, and apparently no organisation was responsible, and that, most serious of all, this had already been made clear in an internal report of the health authority (DIGESA 1984). This found that, in Chimbote for example, there were 15 operational wells and a surface supply. Only two wells were properly chlorinated. Likewise, the Chimbote treatment plant, completely equipped, was not operating properly because of a design and construction deficiency, inappropriate operation and the lack of quality control and records.

All studies of the water supply prompted by the outbreak, confirmed that just before the cholera epidemic, the urban water supply systems had not been operated appropriately in the coastal cities – Trujillo, Chimbote and Lima, among other cities studied – and of the jungle – principally Iquitos – resulting in the production and distribution of poor quality drinking water (see also Alfaro Alvarez et al. 1994, Aurazo 1991, Vásquez 1991). That is to say, that the water was insufficiently disinfected, or simply not disinfected, before its distribution to the population.

In contrast, the drinking water of mountain cities, which was the subject of several tests, was of a notably good quality, even before 1991, in part due to the good quality of the water from its catchment source, but also because disinfection of the water in the treatment plants was carried out in a very acceptable manner.

Other factors have to be added to the deficient environmental treatment conditions, representing the average state of the water supply in urban areas of Peru: the absence of previously acquired immunity to the cholera germ in the Peruvian population; the ability of the bacteria to cause silent infections in many individuals; the prevalence of poor hygiene and sanitary habits in the majority of the population (López Montero 1992).

An environmental factor which has played a major role in the transmission of the disease was the contamination of food products. From the point of view of health, the most dangerous are those produced by street vendors, especially in the street preparation stands and selling to the passer-by consumer, principally in Lima and other large cities. Their direct contamination was because they came in contact with discharges of faeces, be it at the time of handling and storage, or by flies or other insects as carriers.

Perhaps unsurprisingly, the state of health in Peru was also problematic. Around 1986, diarrhoea was the cause of death of 18.2 per cent of children less than 1 year old, and 25.3 per cent of those less than 5 years old. Diarrhoea is steadily on the increase in Peru (from 13.6 per 10 000 inhabitants in 1971 to 127.9 inhabitants in 1990). Diarrhoea constitutes 'the most severe indicator, the inevitable consequence, of the breach between the current provision of sewer services and disposal of excrement and the needs of the population' (Lanata 1991).

The fact that half of Peruvian homes lack plumbing, according to statistics from the Ministry of Housing, shows the level of poverty. Poverty leaves its ominous imprint on the health of those who suffer it. The economic journal, *Cuánto* in November 1990, investigated the impact of the 1990 August currency adjustment on medium- and low-income households. It found that, from June to November 1990, the consumption of calories per capita fell from 2420 to 1962, a decline of 19 per cent of those in the poorest strata.

Poverty caused the low nutritional status of the Peruvian people. On this matter, a project of the population specialist Dr. Graciela Fernández Baca (Fernández Baca 1990) estimated that some 10 million Peruvians suffer from chronic malnutrition. Of these, 3 million are children less than 14 years of age. A more recent study of children from 30 'young towns' around Lima, recorded a considerable increase in cases of children who were underweight, short for their age and verging on acute malnutrition.

Health expenditures have been falling for several years, as was shown by the National Surveys on Standards of Living from July 1985 to July 1990. In this period, households cut their health expenditures by 71 per cent. Health expenditure went from 4 per cent of the family expenditures basket in 1985 to 2 per cent in 1990. Distribution of health services was concentrated in certain cities such as Lima, Arequipa and Trujillo, leaving very few services in the rest of the country, especially Andean towns.

The causes of the origin and the propagation of the disease

At the end of January 1991, the medical authorities in Peru warned of an increase in cases of diarrhoea along the northern coast of Lima. Given that an increase in diarrhoea is normal during the summer season along the coast, initially this was not thought to be an unexpected event.

The first cases of severe diarrhoea were declared in the town of Chancay, a coastal city of 75 000 inhabitants situated some 60 km to the north of Lima (López Montero 1992). By 29 January, 53 proven cases had been declared. In order to discover what was happening, an epidemiological team from the Ministry of Health went that same afternoon to Chancay, to study the outbreak. It was discovered that similar outbreaks occurred almost simultaneously in Chimbote and Piura, coastal cities further north.

The first newspaper reports of the disease also appeared around the fourth week of January, and they referred to Chancay, Chimbote and Piura as the cities initially affected. From these reports it can be concluded that, apparently, the disease had travelled along our northern coast before becoming evident with the first patients. Indeed, a story from the end of January tells of massive diarrhoea in the jail and barracks in Piura (López Montero 1992). Nevertheless, when it was clear that a violent epidemic had begun, everyone turned to Chimbote as the source of the disease.

It was discovered that this was the new cholera on February 13, which had already become a national epidemic. During the eighth week of its advance, the epidemic had arrived in the 23 districts of the Republic, and on July 12 1991, there had been 231 113 cases, of which 89 545 received hospitalisation. Mortality at that date was 2282 dead, a mortality rate of 1 per cent (*Medicación and Salud Popular* 1991).

Juan Rosado Benites, a paediatric physician at Hospital de la Caleta in Chimbote, gave the following statement with regard to the first cases which he witnessed:

> There are normally many cases of diarrhoea among children, but the hospital's physicians have noticed that, during the last week of January, it was cases of diarrhoea among adults that were multiplying day after day.
>
> I could personally see how this disease had caused, on January 31, the death of a woman and one of her grandchildren, while another barely recovered. The hospital's capacity was overrun. The number of patients arriving forced us to attend them in consulting room beds, in the halls, on mats, on benches and even in chairs. They were coming in numbers because they are the ones who don't have enough to go to clinics or to Social Security.
>
> Deaths were more frequent in the first days because patients had to buy their medicine, and we didn't know that in this disease they consume up to 15 litres of polyelectrolyte serums, and it was impossible for many of them to buy it. In these cases, as the patients didn't know the severity of the disease, they arrived in a state of irreversible shock which made it impossible to revive them. No one had the slightest idea what this was because cholera is a disease of poorer countries and poorer people (Reyna 1991).

When the epidemic was discovered, two hypotheses were put forth on the origin of cholera in our country. The first was presented in the magazine programmes *En Persona*, directed by César Hildebrandt. In summary, this version states that on January 11, a ship arrived at Chimbote from China, the *Feng Xian* which moored to load its storage holds with fish flour.

Two sick crew members who had arrived on this ship discharged their faeces, thus infecting the sea at Chimbote. From that time, the disease spread. This theory emphasises the rapidity of the propagation of the vibrio brought by a foreigner arriving only a little earlier on our shores.

Many variations of this hypothesis can be found. Some hold that the disease did not start in Chimbote, but in Piura or in Chancay. Likewise, some state that it was not a Chinese traveller but an

African one, who did not arrive in a ship but by plane, and, while travelling, he spread cholera vibrio down the entire coast.

A second hypothesis states that because it is evident that some of the varieties of the vibrio were already present in Peru, certain favourable conditions, as yet not clearly determined, would have made certain mutations possible, and their massive reproduction favoured the attack on the population, completely unprotected against this disease.

Our hypothesis on the origin and dissemination of the epidemic is that the contamination of the drinking water was one of the most important factors contributing to the spread of cholera in Peru. The failure to disinfect the water in the urban water supply systems in Peru was a critical factor in the propagation of the epidemic. Had effective chlorination of water for human consumption existed prior to and from the very time the existence of cholera in Peru was discovered, the disease would not have spread so rapidly and in such a deadly manner as occurred.

One finding confirming this hypothesis was the high rate of attack of cholera among residents of dwellings connected to the supply system of various cities of the coast and jungle (Haratani and Hernández 1991). Also, this shows that, at least in 1991 the cholera epidemic in Peru was a fundamentally urban phenomenon.

Another finding that confirms this hypothesis is the fact that the epidemic in Trujillo and Chimbote, as well as in other cities studied, was very explosive and not long lived. The rapid increase and sudden fall of the epidemiological curve seen in these cities is typical of an epidemic caused by one single source of infection: the public water system. This being so, these findings support the notion that the measures taken to disinfect the water in these cities, including household disinfection, were effective in preventing an outbreak of a more prolonged duration involving many exposures at different times (Pardón 1991).

The epidemic lasted much longer in cities such as Iquitos and Lima, due to the extreme complexity of the problems resulting from the water supply system of Iquitos and the extremely poor sanitary conditions in which a high percentage of its inhabitants lives. Furthermore, the few cases which occurred in mountain cities indicated that the urban water supply systems which were operated properly – including the catchment and protection of the water supply – constituted a strong barrier against this disease.

The first cases were reported almost simultaneously in three separate cities 400 to 500 kilometres apart. From them, the disease spread to other urban and rural areas of Peru and to many other countries in the Americas. By the end of 1991, 391 220 cases of cholera had been reported in 16 countries in the Americas. Peru, Ecuador and Colombia were the most affected, accounting for 97 per cent of the total number of cases in the region (Ministry of Health 1992).

Although the consumption of fish and raw shellfish was initially blamed as the vector of transmission in Peru, the explosive nature of the outbreaks, particularly in the urban areas of the coast, would suggest a common source for the contagion, as has since been proved.

A peculiar trait of cholera is its proclivity to cause epidemic outbreaks affecting many people at the same time, due to high level of transmissibility (Glass et al. 1982). Several epidemiological studies indicate two principal means of transmission of this disease: through the ingestion of water and food products contaminated with faecal material containing the bacteria and from hydrobiological products coming from environmental reservoirs in estuary waters (semi-salty).

In the first test period performed by the Ministry of Health, the city of Lima was divided into 35 sample areas; 102 tests were performed for chlorine residues and 30 found no residues (Salazar 1993). The bacteriological tests of these 30 points found the presence of coliform organisms at four points. All these were in areas where service interruptions occur. Twelve home tanks were subjected to testing and eight showed the presence of faecal coliform organisms.

If the infection was transmitted over the public water system, then that system was contaminated with faecal residue at some point along the system: at the catchment source, in the plant (when it is not operated properly) or in the distribution system. This happened in the coastal and jungle cities during the 1991 epidemic. The high rates of attack by cholera detected in urban sectors with home connections, and the confirmation of groupings of cases in residential areas serviced by well-water contaminated with faecal residues, confirm this supposition.

Furthermore, another study, performed in September 1991 (see Salazar 1991 in appendix), proved something far more serious: as we can see in their tables in the Appendix, in five of the cities whose water treatments systems were analysed, the chlorination process was irregular, and there was no treatment of used or sewage water. The chlorination process was irregular before the existence of the epidemic, and very far below the minimum value recommended by the Pan-American Health Organization, which is 0.8 to 1.0 mg/l. Chlorine levels were increased in these plants only when the epidemic was established (Cánepa 1991).

Given that cholera was unknown for a century in Peru, none of the population had immunity. Add to this the poor hygiene and sanitation, and it is easy to see that the Cuban specialists, led by Vicente Garcia, were right in including the following in their final report:

> If it is proved in Peru that cases have appeared simultaneously in Chimbote, Chancay, Callao and Piura, or in other places to the south, as has been stated, it would suggest that the *vibrio cholerae* has been in the environment for some time and was probably spread by biological vectors to the Peruvian coast. It is also possible to think about people excreting the vibrio, not causing the illness or producing it only in the weak form in exposed persons.
>
> Now, because favourable conditions exist, we are assuming that the number of *vibrio cholerae* in the environment has increased dramatically, so as to have permitted the appearance of severe clinical cases, which has allowed the disease to be recognised (Garcia 1991).

Government action and the cost of the epidemic

The governmental improvisation which we suffered in 1991 can be summarised in the mordant sentence of the magazine *Caretas* (1991b) which stated that we had beaten the world record in number of cases of cholera: in Africa, during the first month of the epidemic in 1989, there were 35 606 cases and in Asia, 12 785 cases. In Peru, we had, in the first month of the epidemic, 37 538 cases.

The cholera epidemic was a surprise, an unexpected and therefore unforeseen event. Just after the start of the epidemic, the multi-sectoral Commission for the Fight Against Cholera was formed (Ministerial Resolution 1991), which functioned with the complete co-operation and logistical support of the Ministry of Health. It was chaired by Mr. Eduardo Salazar, and comprised representatives of different agencies that, in one way or another, developed health recovery and health education programmes and environmental programmes. The Commission did not only look at problems relating to the environment, but rather it had a full range of functions, dealing with medical, administrative and foreign-aid matters.

However, it soon became clear that there was no unified master plan to control the epidemic and, as was proved subsequently, there was no director officially named in charge of the campaign, neither were there any further meetings of the Commission.

This hypothesis can be confirmed by means of the following facts: (Haratani and Hernández 1991)

- During the cholera epidemic, the office of statistics and information kept a daily update of the status of the outbreak, taking as a basis the real cases declared in each district health office. Unfortunately, due to the resignation of the Minister of Health and the change of high-level officers, the flow of information was blocked in the middle of March. Since when, statistical data on the state of the cholera epidemic were based on estimates.
- The Division of Basic Rural Sanitation, responsible for providing water supply and treatment services to all rural communities with a population of less than 2000 inhabitants, had planned a two-day seminar on basic sanitation, regionalisation and activities related to the cholera epidemic, which was to be held on February 21 and 22. However, at the last moment and without any explanation whatsoever; the Ministry of Health decided not to authorise the seminar, so it was cancelled.
- An effective system for the reporting, recording and correct and timely processing of information about infectious/contagious diseases was never established to the point of complete operation, due to the lack of financial resources.

- The bacteriological control systems for water and foodstuffs were also deficient, due to a lack of minimum financial resources.
- There were serious difficulties in the application of sanitary measures to the sick, both at home and in the hospital. The removal of their excrement became a permanent problem, and no control existed over the hospitals discharging into the public sewage systems or final discharge ports of their issuers. The same occurred in schools, work centres and in other public places.

What can be classified as effective state actions to manage the epidemic up to now have been increased chlorination and monitoring of the existing water supply systems to urban areas and the distribution of chemical products to purify water in the homes, as well as the monitoring of residual chlorine from the treatment units and the distribution system.

Another government decision that should be pointed out is the issuance of a Supreme Decree by which the severity of the cholera epidemic is recognised and which orders the preparation of an emergency plan for the public potable water and sewer sector. The decree charges the Ministry of Housing and Construction with the responsibility of managing this programme, and it authorises it to invest up to 30 per cent of FONAVI (National Housing Fund) funds in potable-water-supply and sewage works, including treatment.

In spite of this, the most serious error by the government at this level was that of not being able to organise a unified and effective command structure to plan, execute and co-ordinate this campaign. This was proved when the required co-ordination for minimum action was lacking among the Ministries of Housing, Agriculture and Fisheries, and local governments and municipalities with respect to the control of foodstuffs, at the beginning of the epidemic. The epidemic was fought almost exclusively in its medical aspects. Prevention was, therefore, incomplete.

From the eighth week of the epidemic, short-term preventative measures were being focused on the urban and metropolitan area centres along the northern coast, where the initial impact of the epidemic was most acute. Nevertheless, it was declared from two sources (Piura and Cajamarca) that the epidemic was not only advancing inland from the coast but that it was also being propagated from the cities to rural areas. Likewise, these reports indicated that mortality among patients in rural areas was higher than among urban patients, and that it could increase to ten times the current level.

Unfortunately, a serious problem, caused by the government itself, was going to cause a new outbreak of the disease. As a result of the epidemic, fishing and all the industries derived from it bore the brunt, because the consumption of fish had been identified as the probable cause of the disease. The Ministry of Health had recommended not eating fish in the form of ceviche (a very popular Peruvian dish, comprising pieces of raw fish marinated in lemon juice), based on international epidemiological experience associating cholera with fish and raw shellfish, as well as recommendations by the WHO. For this reason, the demand for marine products collapsed, and thousands of fishermen, retailers, packers, truckers and workers became unemployed.

According to Ricardo Segura, President of EPSEP, the state fishery of Peru, in January that company generally sold an average of 20 to 30 tons per day of fresh fish and 30 tons of frozen fish, which could now not be measured even in kilograms.

The 'demonising' of fish was echoed in journalistic sensationalism, which found an adequate opportunity to reconcile its most alarming exaggerations with an urgent mission to sound the alarm. Daily headlines with markedly surreal information on cholera frightened the public without informing them.

In the face of this, the reaction of the government in general, and of the fisheries, agriculture, tourism and export sectors, was mutual commercial agreement. Because the government's priority was economic management and within this the strengthening of the treasury, it yielded to the pressures of these groups instead of concerning itself with the health of the people.

This is how what the Peruvian press and politicians called 'the ceviche war' began (which was sparked when President Alberto Fujimori ate ceviche in Pisco on 24 February 1991). While the Ministry of Health was recommending the consumption of cooked fish and shellfish, the Ministry of Fisheries and the President of the Republic himself were making appearances before the press eating raw fish; while the Minister of Health was recommending the consumption only of boiled water and washed fruits, the press was showing the Minister of Agriculture eating unwashed fruit, and the Minister of Housing drinking unboiled water.

This had serious political repercussions. Depriving the Minister of Health of authority, the Ministers of Fisheries and Agriculture, and then the President of the Republic, continued appearing in public, on all the media, eating ceviche and raw fish. They justified these attitudes as trying to calm the export markets, as well as responding to the pressure of businessmen and workers in the affected sectors. This seriously damaged the handling of the epidemic. There ceased to be one single official voice; the authorities contradicted each other, and confusion was created among the public.

After the resignation of the Minister of Health in the middle of March, what was beginning to become a co-ordinated campaign against cholera became a series of unco-ordinated field measures by different governmental entities. Thus an unequalled opportunity for a large health-education campaign was lost.

This tragedy of errors and misunderstandings had detrimental effects. Two days after the President ate ceviche on March 24, hospitals in the Loayza and Cayetano Heredia saw an acute increase in diarrhoea cases, with 66 per cent of cured patients readmitted. A new outbreak of the epidemic, which had been partially controlled, began. The majority of newspapers recorded the new outbreak on 5 March 1991; this was confirmed by the Minister of Health himself on the following day. Asked about the re-outbreak of cholera on 18 March, the President revealed that the Council of Ministers had approved an extraordinary fund of $40 million to meet the problems derived from cholera.

Meanwhile, the theory that Peruvian fish were contaminated was found to be a non-starter, according to the specialists. The cycle of the epidemic, the faecal–oral chain, does not pass through any animal, but is due to the entry of excrement into the inside of an individual, by direct contact.

The recommendation of the Ministry of Health with respect to raw fish and shellfish should have been more specific. As stated by the former Minister of Health, Dr Uriel García Càeceres, 'Ceviche made with fish caught in a contaminated bay is dangerous not because the flesh of the animal is diseased, but because the vibrio is present in the intestines of the fish, which is feeding on it. Then, on removing the guts from the fish, the knife can become contaminated, as well as the chopping boards and the hands of the fisherman or cook. And, then, with those same hands, contaminated ceviche is prepared' (*Caretas* 1991c).

At the economic level, the sectors most affected by the epidemic were tourism, fishing for human consumption in the domestic market, and health. Exports were also affected.

The loss of production can be divided among:

- unrealised sale agreements, that is, lost shipments.
- the reduced prices with which Peruvian products were 'punished' in the international market.
- the increased costs of exportation related to longer storage time in foreign ports due to delays in acceptance on the part of purchasing countries.
- the increased costs of exportation related to stricter quality control to guarantee the absence of cholera.
- technical studies on the prevention of cholera and dissemination which were made by exporters to guarantee safety and improve the image in the international community.
- reduced production caused in sectors which are economically interrelated with exports, due to the losses suffered by the latter (Petrera 1993).

However, it is fundamentally the high incidence of the disease that caused the greatest impact, and this is linked to the standard of living of the population. The care of the patients affected by cholera has implied a reduced supply of other basic health care, which must have been assumed by the households themselves.

The cost of caring for the sick was identified as a direct cost of the epidemic. The public health sector spent $29.05 million in 1991 alone, on the sick persons it could care for (Ministry of Health 1991). Indirect costs refer to the immediate expenses incurred by the State to combat the effects caused by cholera on the sector. This cost relates to the deterioration of the facilities of health establishments, due to the increase in use that they had to sustain.

As the cholera emergency imposed a sudden new demand, other types of care were necessarily severely displaced, because part of the human resources and infrastructure were absorbed by the economy.

150 patients were selected at random from health establishments in the cities of Lima and Chimbote. The results of analysing these patients demonstrated that families affected by cholera differed from those not affected by cholera. They live in housing conditions of greater decay, with deficient basic sanitation and they work under more precarious conditions, principally in the informal sector.

At the end of 1991, Peru had the greatest incidence of cholera, and occupied second place in the worldwide mortality rate in the seventh pandemic. This severe incidence and mortality caused a net loss to the economy of US$489.42 million in 1991 alone. Within the aforementioned loss, the effect of the financial loss from deaths caused is the greatest burden on the economy, resulting in losses of US$406 million (Petrera 1993).

Inasmuch as part of the loss is due to lost future production, only US$255.66 million affect production for 1991; the rest – US$233.76 million – corresponds to future losses in production due to deaths in the economically active and employed population who got sick from cholera in 1991. Toward the end of 1996, Peru's overall losses reached one billion dollars (Merino 1996).

According to the then Minister of Health, Dr Carlos Vidal Layseca, '...unfortunately, [the fact that] this epidemic is affecting precisely the poorest sectors of the country, those which do not have potable water, sewer service and adequate housing, makes us wonder what would have happened in an ultra-liberal government that may have privatised health [services]' (Vidal Layseca 1991).

To correct the Minister, it was precisely the deficient State services that caused the propagation of the epidemic: the extremely low level of acceptance of treatment services and the very poor state of the system prevented the effective control of cholera.

Conclusions

Cholera has revealed the precarious basic treatment conditions in the country with great clarity, and has proved the need to give high priority to the improvement of potable water supply services, sanitary disposal of excrement and sewage water, as well as the hygiene of foodstuffs and the collection and final destination of trash.

All this must be accompanied by health-education campaigns for the population and the institutional strengthening of the national, regional and local agencies responsible for environmental treatment.

The principal measures required in the medium and long term to control the propagation of cholera and other transmissible diseases will require, at a minimum, a large increase in infrastructure investment for water and sanitation supply. An annual capital investment of the order of US$100 to US$120 million is required to provide 80 per cent coverage for the water supply to urban areas and 50 per cent to rural areas, and a sanitation coverage (removal of excrement) of 75 per cent for urban areas and 50 per cent for rural areas by the year 2000 (Haratani and Hernández 1991). This type of investment can only be made by private enterprise.

The disinfection of drinking water continues to be an indispensable measure, not just during the emergency, to protect the health of the population. The possibility of another outbreak of the epidemic will continue to exist if the level of pollution of the coastal and jungle waters increase, along with the lack of an agency that assigns priorities to the use and recovery of water.

Peruvian society has painfully learned that the continuity and maintenance of these measures is a permanent obligation of national, regional and local authorities. If this obligation is abandoned, Peru could once again experience the tragedy it lived through in 1991. If, on the contrary, the educational campaign is retained, the corrective measures taken in the operation and maintenance of the water system are reinforced, and the monitoring of the quality of drinking water is maintained and intensified, Peru will have taken a very important step toward well-being and progress.

References

Alfaro Alvarez, C. et al., (1994). *Cholera Epidemic in Peru*, Lima, Asociación de Consultores Internacionales en Salud.

Aurazo, M. (1991). *Health evaluation of the water supply and treatment system in the city of Chimbote within the framework of the 1991 cholera epidemic*, Chimbote.

Cánepa, L. (1991). *Evaluation of the water supply, sewage and treatment system in marginal urban areas of Iquitos within the framework of the Cholera epidemic*, Iquitos, April.

Caretas (1991a). This article appeared in No. 1236 of the magazine.
Caretas (1991b). This article appeared in No. 1241 of the magazine.
Caretas (1991c). This article appeared in No. 1242 of the magazine.
DIGESA (1984). *Pollution of Coastal Waters in Metropolitan Lima.*
El Comercio (1991). Interview with the Minister of Housing, published in the newspaper on February 12.
El Comercio (1991). These claims and those of the SEDAPAL union, appeared in the newspaper during the third week of March.
Elmore Luján, E. (1991) *The Cholera Epidemic in Peru and its relationship with environmental health problems*, Lima, Pan-American Health Organization, September.
Fernández Baca, G. (1990) 'Appearance and Effect of the Shock', *Cuánto*, Lima, November.
Garcia, V. (1991). *Final report on the first cholera epidemic in Peru and America this Century*, Lima, March.
Glass, P. et al., (1982). *Endemic Cholera in rural Bangladesh 1966–1980*, Washington DC.
Gray, G. M. and Graham, J. D., (1997). Regulating pesticides, pp. 173–192 in Graham, J. D. and Baert Wiener, J., (eds) (1997a) *Risk vs. risk: Tradeoffs in protecting health and the environment*. Cambridge, MA.: Harvard University Press.
Haratani, J. and Hernández, D. (1991). Cholera in Peru: a quick evaluation of the water supply infrastructure and treatment in the country, and their role in the epidemic, Water and Sanitation for Health Project (WASH), No. 331, Field Report prepared by J. Haratani and D. Hernández, Lima, May.
La República (1991). Statements appearing in the newspaper on March 2.
La República (1991). According to a newspaper article dated April 15.
Lanata, C. (1991). *Cholera*, Institute of Health Research, Lima, February.
López Montero, M. (1992). *Cholera in Peru: beginning of an epidemic, hits and misses*, Lima, Pan-American Health Organization.
Medicación and Salud Popular (1991). 'Chronicle: Cholera in Peru 1991', Lima, April.
Merino, B. (1996). Fact mentioned by Congresswoman Beatriz Merino at the conference, 'Environmental Health in Lima: Problems and Solutions', Lima, March.
Ministry of Health, (1991). Ministry of Health's Report to the international Community on the Cholera Epidemic, Lima.
Ministry of Health, General Epidemiology Office (1992). *Cholera in Peru 1991–1992*, which presents statistical information on the

situation of cholera in Peru in 1991 and 1992, Lima, February.
Ministerial Resolution (1991). The Government ordered the creation of a National Multi-sectoral Commission for the Fight against cholera, as per 94-91-PCM, dated April 5 1991.
Petrera, M. (1993). *Economic impact of the 1991 Cholera Epidemic*, Washington, Pan-American Health Organization.
Pan-American Health Organization, (PAHO) (1987). *The Control of Transmissible Diseases in Man*, Scientific Publication No. 507 fourteenth edition, Washington, DC.
Pan-American Health Organization (PAHO) (1991). Epidemiological Bulletin, **12**:1, Washington, DC.
Pan-American Health Organization Mission in Ecuador, (PAHO) (1991). *Guide for the prevention and control of cholera*, Quito, May.
Pardón, M. (1991). *Health Evaluation of the Water Supply and Treatment System in the City of Chimbote within the Framework of the 1991 Cholera Epidemic*, Lima CEPIS/OPS, February 4 to 6.
Reyna, C. (1991). *Chronicle on Cholera in Peru*, Lima, DESCO, April.
Salazar, E. (1993). *Evaluation of the Role of Chlorination in drinking water in the control of the 1991 Cholera Epidemic in Peru*, Lima, January.
Tejada de Rivero, D. (1991). 'The Health Situation in Peru: The Cholera Epidemic', Lima, *Salud Popular* [Public Health], August.
Solórzano, J. (1992). 'Cholera: the version of those affected', Lima, PREDES.
Suematsu, L. (1996). Engineer of the Pan-American Center on Sanitary Engineering and Environmental Sciences, in statements made to the newspaper, *El Comercio*, on March 14.
Vásquez, E. (1991). *Cholera Epidemic in Peru. Case-control study of Piura, February–March 1991*, Lima, Revistade Epidemiologia.
Vidal Layseca, C. (1991). *Cholera and the country where we live*. Conference given at the University of the Pacific, Lima, March.
Waschmuth, I. K., Bopp, C. A., Evins, G. M., et al. (1991). Characteristics of Toxigenic *Vibrio Cholerae* 01 strains isolated from epidemic cholera in South America in 1991. In: *Proceedings of the 27th Joint Conference on Cholera and Related Diarrheal Diseases*, Japan.

Appendix

Table 2.1: Investments in the Andean sub-region in water and sewage in the year 2000 (US$ millions)

	Total investment				
Country	*Water*	*Sewage*	*Overhauling*	*Training and installation development*	*Total*
Bolivia	363	386	76	21	846
Colombia	1477	2114	691	55	4337
Ecuador	700	719	186	40	1645
Peru	1132	1218	866	81	3297
Venezuela	1720	2506	736	108	5070
Total	5392	6943	2555	305	15195

	Annual investment		
Country	*Annual requirement 1986–2000*	*per cent GDP*	*Previous historical investment (annual)*
Bolivia	56	1.6	17
Colombia	289	0.7	142
Ecuador	110	0.9	50
Peru	220	1.0	31
Venezuela	338	0.5	530
Total	1013		770

Source: National Committee on Basic Sanitary Co-ordination, 'Proposal for the Andean Subregion for the continuation of the International Decade of Potable Water Supply and Treatment'

Table 2.2: Principal characteristics of water and sewer services in some Peruvian cities before the cholera epidemic

	Cajamarca	Cajabamba	San Marcos	Iquitos	Trujillo	Chimbote
Geographic area	Mountain	Mountain	Mountain	Jungle	Coast	Coast
Population	91 910	10 800	5000	285 000	630 000	500 000
Water						
Source	River	River	River	River	Well	River and well
Coverage (%)	70	83	90	50	65	50
Other sources	PP, RI	RI	RI	PP, PO, RI, CD	PP, CC, CD	CC
Chlorination	Normal	Irregular	Irregular	Irregular	Irregular	Irregular
Sewers						
Coverage (%)	52	66	90	25	n/a	n/a
Treatment	Yes	No	No	No	No	No

PP = public fountain; CC = tanker truck; PO = well; RI = individual collection along the water flow; CD = covert connection n/a = not available
Source: Salazar, 1991

Table 2.3: Peruvian population (millions)

Area	*1980*	*1985*	*1989*	*1995*	*2000*
Urban	10.2	12.6	14.4	17.4	19.0
Rural	6.6	7.2	7.4	7.7	8.0
Total	16.8	19.7	21.8	25.1	27.0

Source: National Committee on Basic Sanitary Co-ordination, 'Proposal for the Andean Subregion for the continuation of the International Decade of Potable Water Supply and Treatment'

Table 2.4: Real water supply coverage with respect to the 1995 and 2000 goals

		Urban Areas			Rural Areas		
Year	*Total population*	*Population*	*Served*	*per cent*	*Population*	*Served*	*per cent*
1980	16.8	10.2	6.9	68	6.6	1.2	18
1985	19.7	12.6	9.1	73	7.2	1.2	17
1989	21.8	14.4	11.2	78	7.4	1.8	24
1995	25.1	17.4	14.6	84	7.7	3.1	40
2000	27.0	19.1	17.1	80	8.0	4.0	50

Source: National Committee on Basic Sanitary Co-ordination, 'Proposal for the Andean Subregion for the continuation of the International Decade of Potable Water Supply and Treatment'

Using 1989 as a reference, the investment required to achieve the 1995 urban water treatment goal of 69 per cent at a cost per person of US$ 74 would be US $259 million. To achieve the rural treatment goal of 36 per cent at a cost per person of US$35 an investment of US$52.5 million would be required.

Table 2.5: Water supply and treatment needs in Peru (weights in metric tons)

Duration	*Urban*	*Rural*
Short term (8 months)	Current population coverage	Current population coverage
Disinfection only	Water 78%	Water 24%
	Sewer 59%	Latrines 18%
	Water systems 200	Water systems 3000
	Hospitals 100	Latrines 50 000
	Health centres	New latrines 100 000
	Chlorinators 500 units	Total latrines 150 000
	Chlorine gas 2400 tons	Hypochlorinators 3000 units
	HTH crystals (70%) 2000 tons	HTH crystals (70%) 4000 tons
		Lime 40 000 tons
	Estimated cost US$m 4–6	Estimated cost US$m 4–6
Long term	Population coverage goals	Population coverage goals
(1991–2000)	Water 80%	Water 50%
	Sewer 75%	Latrines 50%
	Water $425 million	Water $70.4 million
	Sewer $422 million	Latrines $94.5 million
Total estimated cost	$847 million	$164.9 million

Table 2.6: Accumulated number of reported cases of diarrhoeic diseases, hospitalisations and deaths by health departmental unit Peru, to April 13, 1991

UDES	*Probable cases*	*Hospitalised cases*	*Deaths*
Amazon	107	5	–
Ancash	17 247	5909	48
Apurlmac	2	2	–
Arequipa	3744	654	12
Ayacucho	321	135	24
Cajamarca	6735	3644	326
Huancavelica	6	4	1
Huánuco	241	188	12
Ica	2517	430	10
Junín	547	310	12
La Libenad	25 573	10 644	146
Lanbayeque	12 867	8293	90
Loreto	418	263	18
Madre de Dios	5	–	–
Moquegua	196	81	3
Pasco	10	10	1
Piura	16 483	4230	99
Puno	168	33	4
San Martin	1295	475	65
Tacna	45	20	2
Tumbes	823	551	1
Ucayali	15	15	–
Lima-Callao	54 073	16 865	153
Total	143 438	52 752	1027

Source: OF'S, Epidemiological Bulletin, Vol.12, No. 1, 1991

Table 2.7: Monthly average of residual chlorine measured at different points of the Cajamarca water system

Date		*Average (mg/L)**
1991	October	0.43
	November	0.51
	December	0.50
1992	January	0.48
	February	0.53
	March	0.50
	April	0.50
	May	0.50

* Minimum recommended value according to the Pan-American Health Organization: 0.2–0.5 mg/L

Source: Cajamarca Aid Hospital

Table 2.8: Monthly average of residual chlorine in the Cajamarca water treatment plants 1990–1992

Date		*Concentration of residual chlorine (mg/L)* in*	
		Santa Apolonia	El Milagro
1990	Sept	0.47	0.51
	Oct	0.44	0.52
	Nov	0.40	0.52
	Dec	0.50	0.51
1991	Jan	0.50	0.70
	Feb	0.60	0.70
	Mar	1.00	1.40
	Apr	0.86	1.20
	May	0.65	0.87
	Jun	0.61	0.98
	Jul	0.60	1.00
	Aug	0.58	1.00
	Sept	0.58	1.00
	Oct	0.80	1.20
	Nov	1.00	n/a
	Dec	1.00	n/a
1992	Jan	1.00	1.00
	Feb	0.80	1.00

* Minimum recommended value according to the Pan-American Health Organization: 0.8–1.0 mg/L (n/a not available)

Table 2.9: Monthly changes in residual chlorine [in water] leaving the Iquitos water treatment plant, 1991–1992

Date		*Average Residual Chlorine (mg/L)**
1991		
	January	<0.1
	February	<0.1
	March	<0.1
	April	0.1
	May	0.05
	June	1.0
	July	0.6
	August	0.6
	September	0.9
	October	0.9
	November	0.7
	December	0.6
1992		
	January	0.9
	February	1.0

Source: SEDALORETO

* Minimum recommended value according to the Pan-American Health Organization: 0.8–1.0 mg/L (n/a not available)

Table 2.10: Supporting measurements of residual chlorine tin water leaving the Iquitos water treatment plant, 1991

Date	*Residual chlorine (mg/L)**
23.4.91	0.15†
24.4.91	0.15†
26.4.91	0.6†
5.9.91	1.2
26.9.91	6.0
	0.0

* Minimum recommended value according to the Panamerican Health Organization: 0.8–1.0 mg/L
† Performed by the CEPIS-Commission
Source: Office of Environmental Sanitation of the Iquitos Regional Hospital

Table 2.11: Water quality in the Iquitos system, 1991–1992

Date	*Place*	*Parameter*	*Result*
23.4.91	PJ Micaela Bastidas	CF	NMP 43/100 ml
5.9.91	Different points	CI	<0.5 mg/l
12.9.91	Different points	CI	<0.5 mg/l
19.9.91	Different points	CI	0–2.0 mg/l
26.9.91	Different points	CT	positive
10.3.92	Ichasoles fountain	CT	NMP 10/100 ml

CF: Faecal coliforms
CT: Total coliforms
CI: Residual chlorine
NMP: Most probable number
Source: Office of Environmental Sanitation of the Iquitos Regional Hospital

Table 2.12: Rate of attack by cholera in Trujillo by type of water supply system (1991)

	Domestic connection	*Tanker truck*	*Total*
Population	412 961	217 264	630 225
Cases of cholera	8132	6825	14 957
Rate of attack (%)	1.97	3.14	2.37

Table 2.13: Monthly change in residual chlorine in the Chimbote water distribution system, 1991–1992

Date		*Concentration (mg/L)**
1991		
	January	no measurement taken
	February	0.026
	March	0.043
	April	0.039
	May	0.34
	June	0.30
	July	0.39
	August	0.54
	September	0.57
	October	0.78
	November	0.80
	December	0.90
1992		
	January	0.71
	February	0.73

Source: SEDACHIMBOTE

* Minimum recommended value according to the Pan-American Health Organization: 0.8–1.0 mg/L (n/a not available)

Table 2.14: Results of the analysis of samples of potable water in the city of Chimbote (February 1991)

Sample point	*Residual chlorine (mg/L)*	*Faecal contamination*
Egress treatment plant	0.6	negative
Reservoir FU and R3	0.0	positive
Southern Zone Public System	0.0	positive
Northern Zone Public System	0.0	negative

Source: Canepa, L. et al., 1991

3 A sociology of health panics

Frank Furedi

Summary

In post-industrial societies the standard of health has never been better, life expectancy continues to grow, working and living environments never safer nor better monitored; yet there is a growing obsession with health and safety. Indeed, there has been a 66 per cent rise in self-reported long-term illness in Britain since 1972. The general mood of anxiety about existence makes the public susceptible to health scares, even to the extent of a loss of judgement about risk assessment. People are designated as being 'at risk' not because of their chosen behaviour, but because of who they are. Thus, risk becomes an entity in its own right, only minimally subject to human intervention. The idea of risk has come to have only negative connotations. This paper analyses sociological factors, such as change, uncertainty and crises of confidence, that help to explain why people worry themselves to the point of illness.

Introduction

On both sides of the Atlantic, people are continually obsessed with their health. America's preoccupation with medical issues is so pervasive that most of the time we forget that we are healthier and more prosperous than at any time in human history. According to recent figures released by the National Center for Health Statistics, Americans live longer than ever before and the infant death rate has reached an all time low. Our children are more likely to survive into adulthood than in the past. And despite all the publicity about a range of environmental problems most Americans enjoy a life that is far safer than before.

Paradoxically, the healthier we are, the more we become obsessed with our health. Every aspect of our life is now assessed as a potential health risk. A major study of medical journals in the United Kingdom, Scandinavia and the USA between 1967 and 1991, found a phenomenal increase in the use of the term 'risk'. During the first five years of this period, the number of 'risk' articles published was around 1000 – but for the last five years there were over 80 000. Such a significant increase in the interest in health risk is a testimony to a state of mind rather than the exponential growth of new diseases.

In Britain, more and more people are convinced that they are genuinely ill. According to the recently published *General Household Survey* as many as four in ten people in some parts of Britain consider that they have a long-standing illness. These figures suggest that there has been a 66 per cent rise in self-reported long-term illness since 1972. A report published by the Policy Studies Institute claims that lone parents are particularly prone to suffering from a serious illness. Almost 30 per cent of those interviewed reported a long-term illness. And predictably their children were also suffering from the same condition. Again, the proportion of the respondents reporting children with long-term illnesses was almost 30 per cent. With this explosion in self-reported long-term illness it is possible to conclude that Britain has become a terribly unhealthy, disease-ridden society. Yet nothing can be further from the truth. A baby boy, born in Britain today can expect to live until he is nearly 75; a baby girl till over 80.

It appears that the real health problem in Britain is not physical disease but a general mood of anxiety about existence. Public obsession with health is fed by a regular diet of media coverage of medical research and health news. The most irrelevant bit of health research is guaranteed media coverage, regardless of its coherence and status. Small, obscure studies, which contradict previous findings are reported as serious news. So during the last year publicity has been given to a study that claimed that cot deaths could be caused by pollution from traffic and industry. This report was followed by a study that claimed that quilts put babies at risk of cot death. And this research was succeeded by reports that pointed to the possible association between flying in aeroplanes and cot death. In a sense it does not matter what these reports say – they all confirm public perception of the danger of cot death and strengthen the anxieties of parents.

That these parental anxieties have now acquired epidemic proportions can be seen in relation to the public reaction to media reports in March 1998, regarding the alleged connection between the Measles Mumps Rubella (MMR) vaccine and Crohn's disease and autistic spectrum disorders among children. A relatively insubstantial report was blown out of all proportion, the Government dithered and many parents panicked. By the time the Chief Medical Officer tried to reassure the public that there was no link between MMR and autism, the damage had been done. It was too late to prevent the outbreak of a panic. According to widely cited news reports, during the spring of 1998, as many as one out of four parents were refusing to inoculate their children with the MMR vaccine.

Health panics, whether over the alleged risks of contraceptive pills, vitamin C or over red meat have become an integral part of the British way of life. Concern about health has created a situation where safety has become an important objective in its own right. The search for safety has become a permanent feature of society that has succeeded in turning virtually every object, every type of food and every technology into a health risk. At various times, cars, mobile phones, electric cables, power lines, the Internet, the computer screen or television has been cast into the role of a health risk. Government and public health officials have difficulty resisting the temptation to regularly warn the public about this or that health problem and unwittingly give official sanction to panic-like reactions. At a time when health panics have become banal and when more and more people report that they are ill, it is worth asking, why are we so scared?

The 'at risk' concept

Contemporary discussion of health is most clearly expressed through the conceptualisation of being 'at risk'. This new and original way of framing the term is so pervasive, that it is easy to overlook the fact that thinking of risk in this way is new. To be at risk is an ambiguous concept. It is used to denote groups of people who are particularly vulnerable to a hazard. Children who are at risk are usually associated with a particular lifestyle. It also represents a statement about human beings. Their range of options and their future are circumscribed by the variety of risk factors that effect them. To be at risk also refers to certain situations, encounters and experiences. Sex, family life, communities sited near power stations, or walking out at night are experiences which are said to place people at risk.

The emergence of the at-risk concept ruptures the traditional relationship between individual action and the probability of some hazard. To be at risk is no longer about what you do – it is also about who you are. It becomes a fixed attribute of the individual, like the size of a person's feet or hands. Consequently, experts in different professions draw up profiles of who is at risk. So public health professionals look at the lifestyles of specific communities and claim that this information can be a useful indicator of whether or not their children are at risk. Surveys of risk factors isolate forms of behaviour which are symptomatic of those who are most likely to be at risk. Smoking, obesity, stress are only some of the more publicised risk factors in the field of health promotion. Through mobilising the discourse of risk, public health officials have assumed the right to define what is a good and what is a risky lifestyle. Pregnant women who drink wine, smokers, those who do not keep fit and those who eat beef are considered to be some of the people who are, by definition, at risk.

Being at risk also implies the autonomy of the dangers that people face. Those who are at risk face hazards that are independent of them. It implies that danger is not merely the outcome of any individual act but is something that exists autonomously, quite separate from the actor. Thus, the probability element, where choices about loss and gain informed the decision to take a risk, has given way to an emphasis on avoiding danger. Once risk is seen to exist in its own right and is therefore only minimally subject to human intervention, the most sensible course of action is to avoid it altogether. The diminution of the human agency, that is implicit in the at risk concept has dramatically changed the calculation of risk. The contemporary meaning of risk has little in common with its usage in the past. Openness to the positive as well as negative possibilities of an activity has been overwhelmed by the certainty of adverse outcomes.

If risk is autonomous, it suggests that it exists independently of any act or individual. Like the Greek gods, risk factors exist in a world of their own. The role of society is to warn its members about this complex of hazards with which they are compelled to live. As any risk is a condition of life, the traditional conceptualisation of risk in relation to a specific hazard or technology, is far too limiting. The system of risk factors is represented as being prior to, and independent of any individual act, so the experience of being at risk transcends any particular experience. Attitudes towards personal security express this consciousness no less than reactions to problems of the environment. Consequently, the consciousness of risk influences human behaviour in its totality. The autonomisation of risk factors reverses the human-centred relation between individual and experience. In this scenario, the autonomous individual disappears and returns as one that is subjected to the authority of risk autonomous risk factors.

The perception of being at risk expresses a pervasive mood in society; one that influences action in general. It appears as a free-floating consciousness that attaches (and detaches) itself to a variety of concerns and experiences. The pre-existing disposition to perceive not just major technological innovations but also mundane experiences as potentially threatening means that there is a heightened state of readiness to react to whatever danger is brought to the attention of the public. The mere suggestion that a pill may have some side effect is sufficient to provoke an anxiety attack. And when the object of concern is presented as a really grave threat, the reaction becomes even more intense.

During the past decade, supposed threats to human survival have been declared so frequently that the expectation of an apocalypse has become rather banal. Our imagination continually works towards the worst possible interpretation of events. Expectations of some far-reaching catastrophe are regularly rehearsed in relation to variety of risks. Thus, fears about an explosive epidemic of a lethal infectious disease reinforce anxieties about the dangers of nuclear war, global warming and other environmental disasters. AIDS has retained its status as the modern equivalent of the plague, only to be joined by new threats to health – such as Ebola, Mad Cow Disease and Hong Kong's avian flu – and the re-emergence of old dangers, notably cholera, malaria, tuberculosis and diphtheria, often in drug-resistant forms. The media is continually warning of the creation of a new breed of antibiotic-resistant 'super bug', which cause horrible and fatal infections in hospital patients. Hospitals are no longer seen as places where the ill can get well. On the contrary they have become danger zones where mutant bugs are waiting to infect innocent patients.

We live in a world where medical doomsday scenarios have become routine. Recently published best sellers like Arno Karen's

'Plague's Progress: A Social History of Man and Disease' and Laurie Garrett's 'The Coming Plague: Newly Emerging Diseases in a World Out of Balance', have had a major impact on both sides of the Atlantic. They give coherence to a new strain of panic about plagues and epidemics, which is spreading like wildfire through the worlds of science and popular culture. Never has the word epidemic been used in association with so many different phenomena. So Karen can project a 'massive global die-off', which might result from a 'revived bubonic/pneumonic plague, a virulent new flu virus, a new airborne haemorrhage fever, or germs that lurk undiscovered in other species'. In fact, one of the main objections to proceeding with primate-to-human organ transplant is the apparent risk of transferring unknown viruses from animals to humans. So while there are questions about which disease will threaten human survival – the existence of such a threat is not under discussion. In this scenario, plagues are waiting to be discovered by our free-floating anticipation of danger.

This free-floating anticipation of danger was most clearly expressed in a recent report published in April 1998 by the House of Lords science and technology committee. This alarmist report warned of the danger of new breeds of bacteria, such as methicillin-resistant staphylococcus aureus, (MRSA) known as Superstaph, which are resistant to the existing types of antibiotics. According to the report, unless there is a substantial reduction in the use of antibiotics, it would be only a matter of time before we returned to the bad old days of incurable diseases.

In the decade since the AIDS panic first swept the Western world, there have been a series of dramatic encounters with infectious diseases. Some have been associated with contaminated foods (eggs with salmonella, soft cheeses with listeria) and others have emerged from exotic foreign locations (such as the Ebola outbreak in Zaire). Others still are old-fashioned diseases like tuberculosis and diphtheria. The most recent large scale public health scare in Britain erupted in March 1996 in response to the fear that beef infected with bovine encephalopathy (BSE) has led to cases of Creutzfeldt-Jakob disease in humans.

The most common feature of these disease scares is the systematic exaggeration of the scale of the threat. Infectious diseases – even the most appalling example, AIDS – pose less of a threat than comparable conditions in the past. Take the plague organism itself.

Yersinia pestis killed one third of the population of northern Europe in the four-year period between 1346 and 1350. And those who are enthralled by the uniqueness of the HIV virus, should consider the influenza strain that killed 20 million worldwide in the winter of 1918–1919, more than the First World War itself. And what about the highly publicised super viruses? The much-hyped new viruses – Ebola, Lassa, Marburg and various other insect and rodent-borne bugs – are indeed highly lethal, but as many commentators have pointed out, this renders them less likely to cause an epidemic. They kill their victims too rapidly, before they have a chance to transmit the infection. Hence these viruses tend to cause small and short-lived outbreaks, affecting relatively few people. It has also been widely noted that many more people died unnoticed in Zaire in 1995 from malaria, measles and diarrhoea than the 245 who succumbed to Ebola in the gaze of the world's media. And in Britain the 29 000 deaths due to influenza during the epidemic year 1989–90 also passed comparatively unnoticed.

Scares about infectious plagues are complemented by the continuous discovery of new health problems. Along with epidemic the term syndrome is one of the most overused concepts of the nineties. An increasing range of experience invites an association with a syndrome. We have everything from the Gulf War Syndrome to chronic lateness syndrome. Recently, a woman claimed that she was struck down by a stroke while she had her hair washed in a beauty salon. Promptly, the term 'Beauty Parlour Syndrome' was coined. Panics about futuristic high profile epidemics compete with new lifestyle syndromes for the attention of the public.

Why do we panic?

In sociological discussions, the growth of risk consciousness is blamed on technical factors – such as the massive growth of science and technology. Such explanations are based on the assumption that not only are the long-term consequences of human action incalculable today, but are also beyond control in the future. This objectification of danger renders the human response of panic and fear self evident. It concludes that we are right to worry about the unintended consequences of our actions.

To treat the perception of risk as an understandable reaction to technological development is to underestimate the crucial social processes that it expresses. The negative representation of risk and its relentless inflation, does not take place in a vacuum. People's concern about their health has always been an expression of how they perceive their existence. Our inflated consciousness of risk in general, and health risk in particular are shaped by some of the following influences.

1. *Change is often experienced as risk*
   Perceptions of risk are influenced by the previous experience of social and political change. The failure of numerous social experiments – from that of the Soviet Union to European type welfare state – has strengthened suspicions about the consequence of social experimentation. Today, terms like planning, social engineering and reform, often have a negative connotation. Such reactions pertain not only to political experimentation, as human intervention itself is often seen as mainly destructive and rarely constructive. Initiatives in the field of science and technology are regarded with scepticism that is matched by the certainty that something will go wrong. The fear of side effects, in relation to scientific and technological innovation is the clearest manifestation of this association between change and danger.

   Scepticism towards change indicates that belief in finding solutions to the problems facing humanity lacks plausibility. The discrediting of solutions has gone furthest in the sphere of politics but it has spread to all fields of social engagement. As solutions appear to lose their relevance for our lives, problems assume an overwhelming form. The absence of obvious solutions endows problems with extra weight and importance. The inflation of problems that is a characteristic feature of today's risk calculus follows logically from the decline of support for the perspective of social change. The mismatch between the human endeavour to discover solutions to serious problems in the past is recast as a warning to those who would seek change in the future. The main legacy of the acknowledgement that society lacks solutions is the consolidation of a culture of uncertainty.
2. *Concern about the future*
   Suspicion about change inexorably influences the way people regard the future. The underlying expectations

are that the situation is likely to get worse. Most opinion polls confirm that the public regards the future with fear. For the first time since the end of the Second World War, parents expect that life for their children will be worse than it was for them. Such perceptions of the future reflect contemporary anxieties – indeed they project into the future the collective insecurities of society today.

The future is seen as a terrain that bears little relationship to the geography of the present. Since the process of change appears unresponsive to human management – its future direction becomes more and more incomprehensible. Society's estrangement from the process of change is expressed in a future that is so strange as to be unrecognisable. This is most clearly represented in the media – where the future is treated in a way that highlights its dehumanised difference to the present. Science fiction today projects future society as either a wasteland or as a high-tech purgatory. A similar message is enunciated by some theorists of risk. According to the editors of an influential text on the subject, 'the future looks less like the past than ever before and has in some way become very threatening'.

There have been times before when the future was perceived in such negative and anxious terms. What is distinctive about the way that the relationship between the present and the future is constructed is that the future that we dread is the direct result of our action today. This is clearly expressed in the belief that the potential for human destruction is so great that its dreadful results will not become evident until many generations to come. In this way our fear of danger today is compounded by the knowledge that the full extent of the risks facing humanity will only be clear in the indefinite future. This helps to strengthen the perception of risk as unbounded. The riskiness of our actions will not be known until many decades hence. Consequently, our actions not only put people at risk today but also those of generations to come. It is this model of the future that informs the mainstream of ecological thinking. Terms like intergenerational equity and of sustainability suggest that our action should be restrained by consideration of future development.

It is important to note that when the future is deemed to be very threatening, it is present day society that is condemned. For if our actions are likely to have such an impact on the future, then it is we who are responsible for what happens in the period ahead. As Luhman wrote 'more and more of the future apparently comes to depend on decisions taken in the present'. Since our actions are likely to increase the dangers faced by people in the future – the most enlightened strategy is to minimise the risks faced by future generations. That requires that we do as little as possible of anything that is likely to have future consequences.

3. *Impossibility of knowing*

Increasingly, risk is intellectually defined in relation to our inability to know. At issue is more than not knowing but rather the impossibility of knowing. If the consequences of our action for the future is not knowable then the perception of risk is consolidated. The inability of predicting outcomes is often linked to the fast and far-reaching consequences of modern technology. Many observers argue that since the consequences of technological innovations are realised so swiftly, there is simply no time to know or to understand their likely effect. The lack of time is also posited in relation to the long-term effect of actions taken today. Many supporters of the so-called, 'precautionary principle' argue the need for caution on the grounds that by the time the outcome of a particular innovation is understood – processes which will cause damage to generations to come will have been unleashed. According to Luhman, the absence of time required to obtain the necessary information weakens hope in rationality. It is simply not possible to know much about future tendencies of development. Recently, public health officials in Britain have adopted the term 'theoretical risk' to characterise possible dangers that might arise in the future. The banning of beef on the bone was justified on this basis. In April 1998, pregnant women were warned off mercury tooth fillings on the grounds of the theoretical risk they posed to the foetus.

The concept of a theoretical risk relieves a public health official of any responsibility to prove any medical danger. Since everything is theoretically a risk you can never go wrong. Two months later, in June, the Chief Medical Officer, Sir Kenneth Calman struck again, warning pregnant women and breast-feeding mothers not to eat peanuts if family members suffer from a variety of allergies, such as hay fever or asthma.

The association of knowledge with danger, discussed previously is based on a profoundly anti-humanist intellectual outlook. In this model, knowledge and science are both limited in their grasp of truths. But because they set in motion innovations that have unintended effects they also create problems. Not knowing the outcome of our actions strengthens uncertainty and the negative expectations of events. Not knowing and the sentiment that it isn't possible to know weakens the human capacity to take chances. The expectation of negative outcomes is not hospitable to social experiments. And when suspicion of outcomes are so deeply entrenched throughout society, the quality of reactions to new events becomes at least unstable and anxious. Such responses are but a step away from overreactions and panics.

4. *A diminished humanity*
The negative interpretation of society's ability to manage social experimentation and of the claims of knowledge and science are linked to a vision of society where human beings play a rather minor undistinguished role. The very use of the risk discourse signifies a world-view in which technical factors outweigh social ones. It is worth noting that risk analysis developed in relation to the technological domain, and that the growth of risk thinking expresses the spread of technical calculations into the social domain. Concern with probabilities and predictions inherently point towards outcomes that are, to a considerable extent, independent of human action. Currently fashionable models portray a semi-conscious humanity that is desperately attempting to take control of the forces – mainly destructive – that it has created. In this model, technologically-driven processes have the upper hand and

people are reduced to minimising damage and harm. Such a model expresses a powerful statement about the limits of human control.

The representation of humanity as too powerless to repair past damage and too ignorant to shape the future is in wide circulation. The limited role assigned to human subjectivity is most clearly expressed through risk consciousness. Risks are increasingly posited as autonomous forces that are to a large extent beyond human manipulation. Risks have little to do with any individual or with his or her experience. Risks emerge from a variety of factors that render an individual's action more or less risky. The risks are the active agents and people (at risk) are the passive agents in society.

The growth of risk consciousness is proportional to the diminished role assigned to human subjectivity. During the past decade the role of the human species and the human centred world view (humanism) has been subject to a systematic attack from a variety of directions. Political experimentation has been denounced for leading to totalitarianism. Those who uphold the benefits of science and technology for society are often condemned for an irresponsible lack of concern for the planetary ecosystem. Similarly, the affirmation of the superiority of human reason over animal instinct is often attacked as 'speciesism'.

The diminished role assigned to human subjectivity also implies a redefinition of our humanity. During the past decades the elevation of the passive as opposed to the active side of humanity has been paralleled by concern with people's destructive and abusive potential. The risky individual is also the one at risk. Such attitudes breed suspicion and the disposition to panic.

5. *Reconciling limits*

The spread of risk consciousness has influenced the way people make sense of their circumstances. The diminished role assigned to subjectivity is often discussed in terms of a heightened sense of limits.

Heroes are definitely out. The virtues of the nineties are those of caring and suffering. At the level of the individual, these virtues celebrate the respect of limits. Not taking risks is positively advocated. Since people's powerlessness relative to risks is widely affirmed, limited ambition has become increasingly acceptable. Outcomes beyond human control relieve the stigma of failure. The growth of therapeutic strategies, such as counselling, are based on helping people live with experiences that have put them at risk. The emphasis of such strategies on 'self-esteem' help make indistinct the line that divides success from failure. Knowing your limits, 'accepting yourself' is held to be more important than actual outcomes. The separation of responsibility and accountability from action, at least in an inchoate form, is the most destructive accomplishment of the creation of the diminished subject.

Accountability acquires different qualities in a situation where people live a life of being permanently at risk. The limited scope for human action that this situation affords means that most outcomes are outside any individual's hands. Since the situation is so unpredictable, individuals can demonstrate their responsibility only by playing it safe and not putting anyone else at risk.

6 *Crisis of confidence*

The themes discussed above express a mood where problems are inflated and where possible solutions are invariably discounted. Such sentiments influence the discussion of the economy as much as they inform child rearing or education. Not only is there the absence of the elusive 'feel-good factor', every hint of a difficulty has a tendency to become exaggerated. Most people find it difficult to remain confident about the workings of western society.

The clearest expression of society's loss of nerve has been the institutionalisation of intermediaries who are invited to contain the tensions and conflict that inevitably arise from the struggle to survive. This tendency is underwritten by the sentiment that regards people as both unable and unfit to manage their problems. This conviction is highlighted in the recurrent comparison that is drawn between the 'greedy' eighties and the 'caring' nineties. Such comparisons articulate a criticism of individual pursuit of self-interest and an implicit demand for regulation. Although, this standpoint often seems like an enlightened attack on private greed it can also be seen as an invitation to curb the human potential.

Conclusions

The main reason why today's insecurity has created an intense consciousness of risk has to do with the changing relationship between society and the individual. Many observers have commented on the relentless process of individuation that has occurred in recent decades in Western societies. Changing economic conditions have created an insecure labour market, while the transformation of service provision has increasingly shifted responsibility from the state to the individual. The individuation of work and the provision of services have made survival much more of a private matter. As a recent report by Mintel showed, adults in Britain now tend to look at the future with fear. For most adults (61 per cent), health was the greatest worry. This emphasis on health is important. It is through the issue of health, crime and personal security that a peculiarly individuated concern with survival acquires shape.

But of course, the changes in the labour market alone cannot account for the process of individuation. Economic change has been paralleled by the transformation of institutions and relationships throughout society. The decline of participation in political parties and trade unions points to the erosion of traditional forms of solidarity among people. This has been most clear with the demise of traditional working class organisations, churches and political parties. Many mainstream commentators have interpreted this trend through what they call the decline of community. Even a fundamental institution such as the family has not been immune to this process. The changes in family ties and relations have had a deep impact on people's lives. Today, one out of three children is born outside of wedlock. Among those who marry, the rate of divorce is very high. In these circumstances the security of family life is an ideal that is rarely realised.

The mutually reinforcing combination of economic dislocation and the weakening of social institutions has accentuated the tendency for society to fragment. This problem of social cohesion has implications for the daily routine of individuals. Many of the old routines and traditions of life can no longer be taken for granted. Even the role of the family as a system of support is put to question. Under these circumstances, expectations and modes of behaviour inherited from the recent past cannot be effective guides to future action. Relationships between people 30 years ago may not tell us very much about how to negotiate problems today.

The process of individuation is by no means a novel phenomenon. The break-up of communities, old forms of solidarities, the decline of organised religion, geographical mobility and urbanisation are all important elements in the development of capitalism. However, today's individuation is not merely more of the same. In the past, the erosion of institutions took place in conditions where new forms of solidarities were created. So the growth of the private sphere in the nineteenth century coincided with the emergence of co-operatives, trade unions, mass movements and other collective arrangements. Today, the absence of such arrangements is a widely recognised problem. It has led to the flourishing of initiatives that attempt to provide a substitute for wider social networks. The promotion of self-help groups, telephone help lines and counselling are initiatives designed to compensate for the absence of more organic links between individuals.

The relative weakness of institutions that link the individual to other people in society, contributes to an intensification of isolation. The process of individuation enhances the feeling of vulnerability. Many people are literally on their own. Such social isolation enhances the sense of insecurity. Many of society's characteristic obsessions – with health, safety, security are the products of this experience of social isolation. People who fear alone are uniquely vulnerable to panics. That is why there is such a widespread concern with personal health. Unfortunately, successive government policies have only served to reinforce and intensify such public anxieties. Governments and health professionals have not been able to resist the temptation of constantly reminding people about how unhealthy they are. As a result, the institutions of the state help give shape to individual obsessions about health. The state and the media are perpetually peddling yet another health warning, thereby contributing to the consolidation of a climate of fear.

In an era of widespread mistrust of politicians, Governments have shifted their focus towards apparently non-political issues like education, health and lifestyle. Governments regard such issues as important, since it gives them a rare opportunity to connect with the public mood. That is why, increasingly, Governments are in the business of protecting their citizens from risks – theoretical or otherwise – rather than external or internal foes. And these days everyone is subject to that protection whether they like it or not. Regardless of governments' intent, the effect of these policies is to undermine people's belief in themselves. State sponsored professional advice and help threatens to overwhelm individual autonomy and initiative.

References

Beck, U., Giddiness, A. and Lash, S. (1994). *Reflexive Modernisation: Politics, Tradition and aesthetics in the Modern Social Order*, (Cambridge: Polity Press), p. VII.

Department of Health 98/239, 17 June 1998, Press Release 'Health Advice On Peanut Allergy'. Cited in the *Independent*, 16 May 1996.

Karen, A. (1995). *Plague's progress: A Social History of Man and Disease* (New York: Random House) p. 276. Also see Garret, L. (1995) *The Coming Plague: Newly Emerging Diseases in a World out of Balance*, (London: Virago).

Luhman, N. (1993). *Risk: A Sociological Theory*, (New York: Walter).

Skolbekken, J. (1993). 'The Risk Epidemic in Medical Journals', *Social Science and Medicine*, **40**;3:296. See the *Guardian*, 26 March, 1998.

4 Ecological risk: Actual and hypothetical

K. Ya. Kondratyev

Summary

In order to make a proper assessment of ecological risk, attention must focus on the identification of priorities and differentiation between actual and hypothetical ecological threats. There has been much general discussion of such global problems as greenhouse climate warming and total ozone depletion in the atmosphere. Problems of toxic and radioactive pollution of the environment have been treated in more detail, especially from the viewpoint of its influence on human health. Attention is now being drawn to the inadequacy of using linear extrapolation and ignoring the possibility of a threshold level of impacts when making risk assessments. Until the problem of determining threshold levels is addressed, decisions concerning protection from ecological hazards will not be based upon the full information necessary for robust policy.

Introduction

The contemporary scientific and especially popular-science literature (to say nothing of the media) is overloaded with reports of threatening ecological predictions. In 1962, Rachel Carson's book (Carson 1994) describing the gloomy prospects of the effects of pesticides on the environment and man triggered a stormy reaction. At about the same time dynamic and dramatic predictions of catastrophic climate warming, polar ice melting and a worldwide flood started to appear along with further development of an apocalyptic scenario (Bate 1997, Boden 1994, Carson 1994, Dodds 1997, Elsaesser 1992, Felix 1997, Gore 1993, Houghton 1994, Jastrow 1990, Jaworowski 1997, Kondratyev 1996, 1997a, 1997b, 1997c, 1998a, 1998b, Kondratyev et al. 1996a, 1996b, Maduro and Schauerhammer

1992, Van der Sliujs 1997, Weber 1995). Even Mr. A. Gore, the United States vice-president, joined the discussion (Gore 1993), and his monograph provoked quite contradictory responses (see, for instance, Boden 1994, Jastrow 1990, Kondratyev 1996, Weber 1995). Events such as the Second United Nations Conference on Environment and Development (Rio de Janeiro 1992) and the Special Session of the United Nations General Assembly 'Rio + 5' (New York 1997), as well as conferences arranged by the Intergovernmental Panel on Climate Change (IPCC) should have been of primary importance for scientific and politico-economic analysis and characterisation of the global ecological situation. Regrettably, these global-scale forums have failed to come up to expectations.

There is no doubt that the expansion of man's economic activity and the continuing growth of the world's population (Grigoryev and Kondratyev 1997, Kondratyev 1997a,) are causes for concern for global ecology at local, regional and global levels. (Gorshkov 1995, Grigoryev and Kondratyev 1997, Inge-Vechtomov 1997, Isachenko 1997, Kondratyev 1996, Kondratyev et al. 1996a, 1997a, 1998a, Lavrov 1997, Seliverstov 1997, Trofimov 1997). The problem, however, consists in a quantitative assessment of various manifestations of ecodynamics and in the identification of relevant priorities. Unfortunately, there still are many obscure aspects of the problem which are of principal importance, and, in a number of cases, quite a few unjustified radical conclusions.

The problem of global climate change is perhaps particularly symptomatic in this context. Documents adopted by the above-mentioned worldwide forums contain very radical judgements on the problem. They predict a global ecological catastrophe resulting from an anthropogenically-caused enhancement of the atmospheric greenhouse effect due to greenhouse gas emissions. The predictions have been made on the basis of a so-called 'consensus' in the scientific community. With regard to this 'consensus', although the term is in principle unacceptable and inappropriate for the evaluation of scientific truths, one should admit that there is a general agreement in the scientific community concerning acknowledged observation data. This refers, for example, to data on the increase of the global mean surface air temperature (SAT) or the growth of the greenhouse gas (above all, carbon dioxide) concentration in the atmosphere during recent decades. However, even some of the observational data cannot be related to the category of indisputable facts. For instance, a recent paper by Z. Jaworowski (1997) reveals the doubtful nature of the estimates of carbon dioxide concentration growth during the last one hundred years. It appears that at the end of the last century a serious error was made: an arbitrary selection of data was made when determining 'non-perturbed' concentration because the highest concentration values of carbon dioxide were cast aside. Jaworowski reveals the inaccuracy of paleo-variations of the atmospheric composition retrieved from data analysis of the air chemical composition in ice cores.

Data on long-term air temperature trends are very contradictory. For example, Wiin-Nielsen (1997) processed the satellite measurement data on the mean temperature of the lower 9-km layer of the atmosphere and showed that no statistically significant global atmospheric trend had been observed during the last 20 years or so.

Of course, the results mentioned require further discussion. But there is a clear and urgent need for a critical approach to the now popular concept of greenhouse global climate warming, particularly as this concept is used as a basis for practical recommendations concerning reduction of greenhouse gas emissions to the atmosphere, as an absolute condition for the prevention of a global climatic catastrophe. As noted by Kondratyev (1997b), far more well-grounded assessments of a potential climatic risk are required.

Since the range of problems relevant to global climatic risk have been discussed in detail in monographs (Gorshkov 1995, Kondratyev 1998b, Kondratyev et al. 1996a, 1997), the purpose of this review is an analysis of the urgency of local and regional problems connected with environmental contamination and its effects on public health which have been considered, in particular, in a monograph edited by Roger Bate (1997). The fundamental aim is to demonstrate the decisive role of threshold doses of any toxicants from the standpoint of their effects on the biosphere and man, mostly on the basis of the results discussed in the monograph (Bate 1997).

Toxins and the problem of dose threshold levels

Carcinogens

As has been noted (Wilson 1997), the current practice of estimating the danger of carcinogens in the United States is based on a 'no-threshold' policy, i.e. it does not take into account the threshold concentration values determining the presence or absence of danger. This kind of practice may have been acceptable a few decades ago; but nowadays a more precise determination of the risk of cancer is possible by taking into consideration two biological processes responsible for the risk. One is the increase of the rate at which cells divide (the mitotic rate), the other is the increase of the rate at which mutations occur, which is independent of the mitotic rate. Correspondingly, two types of carcinogen may be identified:

1) mitogenic carcinogens which act predominantly by increasing the rate at which cells divide (increasing the probability of spontaneous mutation) and
2) mutagenic carcinogens which act directly on DNA (causing errors in replication).

Mitotic rate is under close physiological control, operated through a complex system including a variety of intercellular messenger molecules, thereby ensuring a relatively stable rate of cell division. Functions so controlled remain within certain limits in the face of external stressers (such as potential carcinogens). They must exhibit a threshold in their response to small changes in such external stress. So long as the external stress induced by a mitogenic carcinogen is within the limits of the physiological control system, the mitotic rate will not change. All mitogenic carcinogens exhibit a threshold, the level being determined by the individual's physiological response. Therefore, the low-concentration mitogenic carcinogens should be considered as conventional toxicants.

The weight of evidence favours the conclusion that classical mutagenesis is characterised by the presence of threshold perturbations since the rate of mutations is controlled by physiological processes. The notion of a 'threshold' is ambiguous enough, but the concept of a 'practical threshold' is becoming broadly recognised (which implies that no consequences can be detected of effects that are below the threshold level). The use of the concept of the 'practical threshold' makes it possible to avoid some variance inherent in the 'no-threshold' concept. For instance, as it is a fact that most foods contain minor amounts of naturally occurring carcinogens, banning all carcinogen-containing foods would mean forbidding the consumption of most food products. A similar situation took place in the USA after the use of asbestos had been completely forbidden by the Delaney Clause (Wilson 1997). The important task now facing scientists is the development of numerical modelling methods to aid the establishment of criteria or indices which characterise the potential carcinogenicity of various chemical compounds, their threshold levels of dangerous concentration, and also relevant doses.

Benzene and leukaemia

A well-known example illustrating the problem of ecological risk is the problem of benzene emissions (mostly from vehicle exhaust) and leukaemia, which is often considered to be a consequence of such emissions. This problem became evident for the first time in the 1970s at shoe factories in Turkey and Italy, where workers were exposed to intensive benzene emissions, and some of them became ill with leukaemia. Nowadays the cause and effect of the case is a generally adopted reality; however, some of the mechanisms of this link are not clear enough (Bate 1997). Measures have been taken to protect the workers, as much as possible, from the effect of benzene and, consequently, the risk of getting leukaemia. Benzene concentration limits have also been established (though varying within a very broad range in different countries: from 0.005 to 10 million^{-1}).

According to the data of the World Health Organization (WHO), the worldwide production of benzene, mostly due to crude oil processing and vehicular exhaust emissions, is about 14.8 million tonnes per year, which rises to 30 million tonnes if the benzene in fossil fuels is included. A considerable proportion of this is due to benzene's use as an industrial solvent in many manufacturing processes, such as shoe making, printing, dyeing, painting, the production of artificial leather, linoleum, car tyres, among others. Benzene is emitted to the atmosphere from natural vegetation burning. Tobacco also contains appreciable concentrations of benzene. The non-industrial impact of benzene on humans is practically unavoidable. Assessments of the risk of leukaemia were made, in particular, by way of linear extrapolation of the results of examination of workers occupied in the production of synthetic rubber (automobile tyres). However, as revealed by Munby and Weetman (1997), the results proved to be overestimated because the linear extrapolation was arbitrary and based on inadmissible assumptions; hence the overestimation of the effect of small doses. Analysis of available information has revealed that for the risk involved in the impact of small non-industrial amounts of benzene, there is a certain threshold level below which such an impact is harmless. Since this level is hardly ever reached, it may be assumed that the risk of leukaemia due to non-industrial effects of benzene is practically equal to zero. Taking into account the insufficiency of relevant information, it may be assumed that the risk to humans due to benzene present in the atmosphere is either very insignificant or is totally absent.

Tobacco smoke and lung cancer

It is biologically plausible that environmental tobacco smoke (ETS) has a contributory role in the induction of lung cancer in non-smoking individuals. However, recent findings suggest that a major part of the observed increase in lung cancer risk for ETS-exposed non-smokers is the result of misclassification of smoking status, 'controls' not properly matched to 'cases' and ignoring certain factors related to hereditary disposition and lifestyle.

Misclassification is inevitable when smoking status is self-reported in surveys. Data on past smoking habits given by proxy respondents, for example, the spouse of a dead smoker, are inevitably unreliable. Verifying self-reported smoking status is also problematic as there is no satisfactory biomarker for tobacco use. Even where researchers took samples to be tested for the presence of cotinine (a breakdown product of nicotine), the amounts present will vary in individuals, due to differences in the rate of metabolism of nicotine. Further, due to the short half-life of cotinine, a self-reported non-smoker may, in principle, have been a life-long heavy smoker until just before sampling took place.

Some major studies included an excessive number of disease-prone individuals, mostly of low socio-economic status, as cases. This apparently led to a disproportionately large effect on the overall lung cancer rate due to a high intake of saturated fat, coupled with an inadequate intake of anti-carcinogens present in fruits and vegetables.

A consequence of the linear extrapolation of the results of examinations from high ETS doses (active smokers) to low ones, ignoring any threshold level of safe exposure, is the overestimation of the lung cancer risk in non-smokers by at least ten-fold (Nilsson 1997). Despite these findings, the anti-tobacco lobby has gained such political momentum, and the danger of ETS already assumed as fact, that coercive legal action will be taken even without the scientific support on which it must rest. Further, the one-sided preoccupation with ETS as a causative factor of lung cancer in non-smokers may be seriously hindering the elucidation of the true and various causes of the disease.

Pollution, pesticides and cancer misconceptions

As noted by Ames and Gold (Bate 1997), the major causes of cancer are:

1) Smoking, which is responsible (according to data for the US) for about a third of cancer and 90 per cent of lung cancer;
2) Dietary imbalances, for example, lack of dietary fruits and vegetables: the quarter of the population eating the least fruits and vegetables has double the cancer rate for most types of cancer compared with the quarter eating the most;
3) Chronic infections (mostly in developing countries);
4) Hormonal factors influenced by lifestyle.

In the context of the problem of cancer, numerous misconceptions and myths have appeared, among which the following are most essential:

1) **The statement that cancer rates are soaring**. In reality, no cancer epidemic has ever been registered, except lung cancer due to smoking. In the US, cancer death rates have declined 15 per cent since 1950 (cancer is one of the degenerative diseases of old age, increasing exponentially with age; average life duration in the US has been increasing).

2) **The statement that environmental synthetic chemicals are an important cause of human cancer.** Neither epidemiology nor toxicology supports the idea that synthetic industrial chemicals are important for human cancer (in comparison with the above-mentioned factors of influence). Although some experts have come to the conclusion about an existing link between cancer and industrial pollution of the environment, the link is weak, and the results of investigations rather contradictory.
3) **Overestimation of harm from pesticides.** Fruits and vegetables are of major importance for reducing cancer. If they become more expensive by reducing the use of synthetic pesticides, cancer is likely to increase because fruits and vegetables contain micronutrients which are anti-carcinogens. Other micronutrients are likely to play a significant role in the prevention and repair of DNA damage (e.g. vitamin B12).
4) **The idea that human exposures to carcinogens and other potential hazards are nearly all exposures to synthetic chemicals.** On the contrary, 99.9 per cent of the chemicals humans ingest are natural. The amounts of synthetic pesticide residues in plant foods are insignificant compared with the amount of natural pesticides produced by plants to defend themselves from various predators (of all dietary pesticides that humans eat, 99.9 per cent are natural). On average, Americans ingest 5000 to 10 000 different natural pesticides and their breakdown products, i.e. about 1.5 g of natural pesticides per person per day, which is about 10 000 times more than they consume of synthetic pesticide residues. Even though only a small proportion of natural pesticides has been tested for carcinogenicity, half of those tested are rodent carcinogens. Cooking foods produces about 2 g per person per day of burnt material that contains many rodent carcinogens and many mutagens, whereas the residues of 200 synthetic chemicals (including pesticides) average only about 0.09 mg. The known natural rodent carcinogens in a single cup of coffee are about equal in weight to an entire year's worth of carcinogenic synthetic pesticide residues (it is important

to note that only 3 per cent of the natural chemicals in roasted coffee have been tested for carcinogenicity). This does not mean, however, that coffee is dangerous; it only means that cancer risk assessments performed for high-dose animal cancer tests cannot be extrapolated for estimating human risk at low doses.

5) **The assumption of a great cancer risk for humans from synthetic chemicals in comparison with natural compounds, based on the above-mentioned extrapolation.** In reality, cancer risk from synthetic pesticides is low in comparison with that from natural pesticides (both are harmless in low doses).

6) **The hypothesis that the toxicology of synthetic chemicals is different from that of natural chemicals.** It is often assumed that because natural chemicals are part of human evolutionary history, whereas synthetic chemicals are recent, the mechanisms that have evolved in animals to cope with the toxicity of natural chemicals will fail to protect against synthetic chemicals. This assumption is flawed for several reasons, among which there are (Ames and Gold 1997): a broad enough and universal nature of human protection mechanisms; the carcinogenicity of some natural toxins; the imperfection of human evolution from the viewpoint of achieving a 'toxic harmony' (it should be remembered here how fast the human diet has changed in the last few thousand years); overestimation of danger from DDT. The latter is due to the over-estimation of the hormonal dangerousness of DDT and other synthetic compounds (against the background of the underestimation of a similar role of natural chemicals).

7) **There is no risk-free world; hence the problem of key importance is that of priorities, especially because resources are limited and risk regulation measures are extremely expensive (in the USA relevant expenses amount to $140 billion per year).** This is a multi-aspect socio-economic problem that should be resolved in the context of the general problem of human health and reasonable lifestyle. It is in this context that measures for regulating carcinogens and providing the necessary standards of environmental quality should be substantiated. One-sided approaches are doubtless counter-productive.

Dioxins

Dioxins (chlorine-containing organic compounds) are toxic by-products which are generated in small amounts by natural burning processes and technical synthesis of some chlorinated organic compounds. Dioxins have often been considered as the most dangerous toxic substances (Müller 1997). Dioxins were synthesised for the first time in the last century, but the first observations of dioxin toxicity did not appear until the late 1940s following accidents at several herbicide manufacturers (the first of which took place in Monsanto, USA, in 1949). At first, no serious consequences of the effect of the most toxic dioxin, the 2,3,7,8-tetrachlorodibenzo-p-dioxin (TCDD), were noted (a single death during three decades), but by the late 1970s the public opinion had accepted the idea, peddled by environmentalists (though scientifically unjustified), that dioxins are extremely dangerous.

The 1976 Seveso plant accident was a landmark case for dioxins for several reasons. On 10 July 1976, an explosion occurred during the production of 2,4,5-trichlorophenol in a factory near Seveso, Italy, about 25 km north of Milan. A cloud of toxic material, estimated to contain between 159 g and 2 kg of TCDD (estimates differ), escaped into the environment and debris fell on an area of about 2.8 km^2. People soon detected skin lesions and, seeing many animals (including birds, chickens, and rabbits) dying and trees losing their leaves, they were frightened. The authorities divided the contaminated area into three zones (A, B, and R) and, within 20 days, evacuated 735 people from zone A. There were 4699 people living in zone B, and 31 800 in zone R. All the residents underwent extensive medical examinations from 1976 to 1985 which revealed minor symptoms of affection in a small segment of the population.

However, the Seveso accident became internationally known and it also became a symbol of a catastrophic danger for ecological movements and a reason for banning the production of chlorine and all organic compounds containing it. The accident stimulated research to examine the toxicity of TCDD and other dioxins. In order to attract research funding, the impacts of dioxins were over-stated and TCDD became known as 'the most toxic man-made chemical known'. It is true that a few dioxins are very toxic, but only to the most susceptible animals (for example, guinea pigs are a thousand times more susceptible to dioxins than humans).

Food is the main source of dioxins for man, but their concentration in food products is still 50 to 100 times less than the threshold value (which determines the level of a dangerous impact). Since the banning of the production of dioxins resulted in a decrease of their concentration in food products, apprehensions concerning the threat of dioxin poisoning cannot be regarded as scientifically justified. The search for adverse health effects from dioxin has found no evidence sufficient to link dioxin to human cancers, suppression of immune function, or reproductivity.

The problem of ionising radiation

Since the time of the Chernobyl catastrophe, the problem of the effect of ionising radiation on man and nature has been attracting special attention. Z. Jaworowski (1997a) performed an accurate analysis of the problem. It appears that, contrary to the prevailing wisdom, human exposure to ionising radiation – from whatever source, be it nuclear fallout or x-ray diagnosis – exhibits a threshold effect: below a certain exposure level, impacts are no longer detrimental. Moreover, significant evidence supports the hypothesis that low doses of radiation are actually beneficial. This is known as the 'hormetic effect'.

Since 1959, the regulation of admissible radiation doses has been determined by the documents of the International Commission on Radiological Protection (ICRP), based on an assumption that the impact of low dose radiation may be derived by linear extrapolation from high dose radiation. Jaworowski (1997a) remarks that the absurdity of this assumption can be seen by considering the impact of exposing individuals to 10 000 millisieverts (mSv) of radiation. Such a dose would be likely to cause death in a matter of hours. On the other hand, extrapolation would imply that 1 mSv would result in the death of one out of every 10 000 people exposed. This is ridiculous because 1 mSv might cause one DNA lesion in one cell in the human body, but natural, spontaneous DNA lesions occur at a rate of about 70 million per cell per year. If we can survive this relatively high rate of spontaneous DNA lesions, there must be mechanisms that are capable of combating a minor harm caused by 1 mSv of radiation.

Dramatic radiation hormesis was observed in 1943, during the early days of the Manhattan Project, when animals were exposed to uranium dust at levels that were expected to be fatal. Instead, the animals lived longer, appeared healthier and had more offspring than the non-contaminated controls. Despite this and other similar facts, until very recently, reports by national and international organisations, including the United Nations Scientific Committee on the Effects of Atomic Radiation (UNSCEAR) and the American National Research Council Committee on the Biological Effects of Ionizing Radiation (BEIR), consistently failed to acknowledge the hormetic effect at low doses. A similar effect has been seen in the findings of several studies investigating radon gas, which strongly indicate that radon levels in homes are inversely correlated with incidence of lung cancer. If there was a linear relationship between dose of radiation and effect, the result would be the opposite.

The linear hypothesis assumes that harmful post-irradiation effects, such as neoplasms and genetic diseases, appear not only after high doses but also after extremely low doses of radiation, and only that the frequency of the effects is proportional to the dose. According to this hypothesis, the effects of irradiation can only be negative, such as occurrence of cancers, genetic damage, and decrease in life duration. The hypothesis implies no new effects at low doses. One of the consequences of this hypothesis was the development of the sievert unit that is supposed to characterise the risk, independently of the dose. Another consequence was the concept of collective dose, according to which it is possible to aggregate negligibly small individual doses received by whole populations of particular regions, or even the entire world, to frightening imagined levels. The hypothesis justifies adding the doses absorbed by a person at various times and in various parts of the body, without taking into account the duration of periods between particular doses. It also allows aggregation of the doses received by any number of persons in current and future generations.

The linear hypothesis was accepted in 1959 by the International Commission on Radiological Protection (ICRP) as the philosophical basis for radiological protection. Since its very beginning, in 1955, UNSCEAR gained rather a privileged political position, reporting directly to the UN General Assembly, and over the years became the most distinguished international scientific body on matters of ionising radiation. In 1958 UNSCEAR stated that contamination of the environment by nuclear explosions would increase radiation levels all over the world. However, two important points ought to be noted:

1) The 'linearity' hypothesis has been assumed for purposes of simplicity;
2) There may or may not be a threshold dose.

Both possibilities have been retained.

The latter means that the original UNSCEAR view on linearity and threshold remained ambivalent, and this 'dualism' has been kept because of radical differences in assessments of irradiation consequences in accordance with various assumptions about the specific nature of the impact. For example, continuation of nuclear weapon tests in the atmosphere was estimated to cause up to 60 000 leukaemia cases world-wide if no threshold is assumed, and zero leukaemia cases if a threshold of 4000 mSv exists.

Although the linearity hypothesis had not been properly substantiated scientifically, it became very popular and regarded as such by the media, the general public and even individual specialists. In the 1970s ICRP extended the no-threshold principle to exposure of the general population to man-made radiation, and in 1980s it extended the principle to limiting the exposure to natural sources of radiation. The dose limit for the public was set at 50 mSv over a lifetime. This value is less than one-third of the global average lifetime dose from background radiation (168 mSv) and many tens or hundreds of times lower than the lifetime dose in many regions of the world. Limiting exposure below the levels of natural radiation is a logical consequence of the administrative assumption from 1959; if each dose is detrimental, then one should also attempt to decrease the risk of natural background radiation even at such trivial doses as 1 mSv per year. The limit of 1 mSv per year, recommended by ICRP, was supposed to protect the general population against the appearance of so-called stochastic effects (neoplastic and genetic diseases) caused by radiation damage of DNA.

The absurdity of the linearity principle was brought to light by the Chernobyl accident in 1986. On the basis of this principle and following the ICRP recommendations, about 400 000 people were resettled in Belarus, Ukraine and Russia. According to Jaworowski, these people are actually nothing but 'linearity victims' (Jaworowski 1997a, 1997b). Resettlement caused great suffering to its victims. The material losses were estimated at tens of billions of dollars. It has been estimated that in the impoverished Belarus alone the costs of resettlement will amount to US$ 91.4 billion by the year 2015. The intervention level for evacuation was first set as a 70-year lifetime radiation dose of 350 mSv, or about twice the worldwide average natural background dose. In 1991, five years after the relocation of the first 135 000 persons, the Supreme Soviet of the USSR lowered this level to 70 mSv. However, no increase in cancer death rate has ever been observed after irradiation with such small doses received over 70-years lifetime, nor has any such increase been observed after natural doses tens and even hundreds of times higher. In this context one fails to understand why, for example, is not everyone evacuated from Norway, where the average lifetime dose is 365 mSv and in some districts 1500 mSv. Should not regions of India with average lifetime dose of 2000 mSv and Iran with more than 3000 mSv be depopulated?

With regard to low doses of ionising radiation, abundant data have been accumulated which reveal their beneficial and protective effects. For instance, there is evidence of increased growth rate in bluegreen algae exposed to x-rays. The results of some experiments show longer survival times of mice and guinea pigs exposed to small doses of gamma radiation and fast neutrons. Although we are far from understanding fully the nature of hormetic processes and mechanisms, here are some of their manifestations: stimulation of DNA repair, protein synthesis, gene activation, production of stress proteins, detoxification of free radicals, activation of membrane receptors and the release of growth factors, stimulation of immune system, and others. The hormetic effects were found at biochemical, cellular, and organic levels, in cell cultures, bacteria, plants, and animals. In mammals, radiation hormesis enhances defence reactions against neoplastic and infectious diseases, increases longevity, and improves fertility. For example, in an experiment with mice the incidence of leukaemia, cancers and sarcomas was lower in animals irradiated with cesium-137 gamma radiation doses of 2.5 to 20 mSv than it was in non-irradiated controls.

All this indicates that ionising radiation may be essential for life. Living organisms developed under constant exposure to ionising radiation, which 3.5 billion years ago was about 3 times higher than now. Our defence mechanisms cope with the extremes of natural radiation levels: from 1 to 376 mSv/year which is greater than the range of normal exposure to heat covering about 50°C. For example, increasing the water temperature in a bath by a factor of 1.27 from 293K to 373K may cause death, but a single lethal dose of ionising radiation delivered in about an hour, which for man is 3000 to 5000 mSv, is more than 10 million times higher than the average natural radiation dose received during that time.

The existence of the effect of radiative hormesis (and thus the rejection of the linearity hypothesis) has been confirmed by the results of medical examinations of the nuclear attack survivors from Hiroshima and Nagasaki. It appears that among those who received doses lower than 200 mSv, there was no increase in the normal number of total cancer deaths. In fact, mortality caused by leukaemia was lower in this population at doses below 100 mSv than among the non-irradiated inhabitants of these Japanese cities. The Hiroshima and Nagasaki data indicate that a single irradiation with doses between 400 and 600 mSv did not cause detrimental effects in the next generation. Unexpectedly, rather positive effects appeared.

A similar study was carried out more recently in China. Between 1970 and 1986, 74 000 people in Yangjiang county, which has a high level of natural background radiation (5.5 mSv per year), were compared with 77 000 people in two adjacent low-background counties (Enping and Taishan, 2.1 mSv per year). In the high-background Yangjiang county, the inhabitants receive a 70-year lifetime dose of 385 mSv, which is higher than the intervention level for evacuation adopted for Chernobyl, and 5.5 times higher than the dose limit recommended for a population by ICRP, and now implemented in the regulations of several countries as well as the European Union. In an age group of 10 to 79 years the general (non-leukaemia) cancer mortality was 14.6 per cent lower in the high-background county than in the low-background ones. The leukaemia mortality was also lower in Yangjiang. No difference in the frequency of various genetic diseases was noticed between the counties. There are interesting data on mortality in Canadian, American and British nuclear industries for people occupationally exposed to low radiation doses (on the average,

49 mSv for men and 5.5 mSv for women): a lower mortality from leukaemia and other cancers has been revealed.

So, the results of various studies discussed here, and many others, show that it is necessary to take into account the hormetic effect to avoid overestimation of potential risks from ionising radiation.

Decision-making

The above-discussed examples reveal the fact that information, relevant to various aspects of ecodynamics induced by anthropogenic impacts and used by authorities, is frequently insufficient for decision-making. This can be related to changes in the environment (climate, ozone), consequences of nuclear catastrophes (particularly the Chernobyl accident), and ecologically-induced (environmental pollution, pesticides, for example) diseases. Very often, governmental and other bodies responsible for taking decisions, face the necessity of making their decisions under conditions of uncertainty: hence the tendency towards overestimation of potential danger.

In this context, two cases considered by Everest (1997) are of interest: the gradual elimination of leaded gasoline and the control of sulphur emissions in Western Europe. In the first case, the scientific evidence suggested that lead in petrol was not responsible for the apparent decline of intelligence in some children. In the second case, the evidence did not suggest that sulphur dioxide from Britain was responsible for a significant proportion of the acidification of Scandinavian lakes.

An estimate of possible effects of moving to lead-free petrol was obtained by the study undertaken of the effects on air lead and blood lead concentrations of reducing petrol lead from 0.4 to 0.15 g/l that occurred in 1986. This reduction in the lead content was greater than that obtained by going from 0.15 g/l to lead-free petrol. It resulted in a 50 per cent reduction in air lead concentrations, and was accompanied by a 16 per cent decrease in blood levels for children and 9–10 per cent decrease for adults. However, these reductions must be considered in the context of a continuing long-term fall in blood lead concentrations of about 4–5 per cent a year, presumably from other pathways (with food and water) by which lead reaches adults and children (as a rule, the contribution of petrol does not exceed 20 per cent). This downward trend is probably more significant for reducing lead body burden than the expensive benefit achieved by removing lead from petrol.

During the 1970s and 1980s, concerns were growing over the long-range transport of sulphur dioxide and other acid-forming emissions such as nitrogen oxides. Wet deposition (acid rain) of these emissions was supposed to be causing damage to acid-sensitive ecological systems, particular concerns being expressed by the Scandinavian countries where the soils are considered vulnerable to acidification. In 1984 the so-called '30 per cent Club' was formed (originally, an informal group of 10 countries) which pledged to achieve a 30 per cent reduction in their sulphur dioxide emissions by 1993 compared with 1980 levels. (The UK refused to join this club, instead offering to achieve the 30 per cent reduction by the year 2000, but later implemented the recommendations.) However, there is no doubt that scientific efforts in studying the problems of sulphur dioxide emissions and environmental acidification have not been sufficient (for example, their admissible threshold impact on ecosystems has not yet been substantiated). Although priority is usually given to flue gas desulphurisation technology (FGD), the substitution of clean natural gas for coal in power generation is a more acceptable environmental solution to the emission problem. The prospects of further development of nuclear power production should be studied more thoroughly.

Since the estimates of ecological danger are often based on the results of numerical modelling, it is important to understand that the adequacy of models is limited, and that the results of modelling should therefore be further improved by taking into account the data of experiment and theory (which mostly is the case). In this connection, it is essential that the principle of precautionary action should be tempered by taking account of the following considerations:

1) the strength of public concern about the particular environmental issue, and on whom the economic burdens of precautionary action will fall;
2) the differences of view regarding the extent of environmental damage and the strengths of the links between perceived damage and the claimed cause;

3) moral considerations such as the need for equity between present and future generations.

In practice, wider political considerations, often reinforced by apparently strong public perception of environmental risk and damage, usually have greater weight than scientific argumentation in the formation of environmental policy (and sometimes even a 'distortion' of scientific judgement took place).

'Should we trust science?' is a question put by Kealey (1997) who notes that scientists are human, so when they see the possibility of funding, they have an incentive to ensure that their research comes up with the 'correct' answers. A logical consequence of this is the 'distortion' of science, examples of which are, alas, numerous (Inge-Vechtomov et al. 1997, Kondratyev 1996, 1997, 1998a, Jastrow 1990, Maduro and Schauerhammer 1992). This 'distortion' is particularly dangerous if scientific judgement becomes monopolised, as was the case, for instance, with the IPCC Report on 'global warming' (Kondratyev 1997b, 1998a, Jastrow 1990), or with the above-described evaluations of the role of tobacco smoke in causing lung cancer in non-smokers. All these circumstances reveal how necessary is the demand for non-biased scientific judgement, protection of science and the unmasking of irrational technophobia caused not only by real but (possibly to a greater extent) also by imaginary ecological threats, as justly stated by Vahrenholt (1997). It is especially important because 'to some sections of media only bad news is good news' (Bate 1997). And there is always a host of charlatans waiting to join the public debate. One can certainly join Vahrenholt who states (1997): 'It is only of use to society if it finds the courage and patience to distinguish between risks that are really high and those that are just highly discussed, between relevant and negligible input paths, and to make this so transparent that it is possible to form sensible rules of personal behaviour for our daily lives'.

Sandman (1997) is quite correct to emphasise, in this context, the role of the media in the exaggeration of actual ecological risks. Analysis of ample material provided by the media makes it possible to formulate the following conclusions and recommendations:

1) The amount of coverage of an environmental risk topic is not necessarily related to the seriousness of the risk in health terms. Instead, it often relies on traditional journalistic criteria like timeliness and human interest.

2) Most of the coverage is not about the risk, but rather about blame, fear, anger, etc.
3) When technical information about risk is provided in news stories, it has little, if any, impact on the audience.
4) Alarming content about risk is more common than reassuring or moderate coverage.
5) Whether the same piece of information is alarming or reassuring is very much a matter of opinion. As a rule, the media tend to supply alarming information.
6) Reporters lean on official sources on the 'safe' side and use 'activists' and citizens on the 'risky' side when they need them.
7) For several reasons, journalists are more closely connected with 'alarmist' than reassuring sources of information.

Circumstances like those described here lead to four types of 'distortion', or four biases that still prevail: alarm prevails over reassurance; extremes over the middle; opinions over data; outrage over hazard.

Conclusion

It is common knowledge that the continuing population growth on the Earth and the expanding scale of man's economic activity provoke an enhancement of anthropogenic pressure on the environment at every level from the local to the global scale. Numerous manifestations of this enhancement and their consequences have been discussed in detail in the document **Agenda 21** approved by the United Nations Organization. However, there are problems of crucial importance that remain unsolved, viz.: the lack of scientifically substantiated priorities (which is responsible for the failure in the attempt of preparing a conceptual document **The Earth Charter**); and the deficit of adequately detected dangers (risks). The latter is of particular concern now that even among specialists (to say nothing of the media) there doubtless exists (and sometimes prevails) a tendency towards an exaggeration of risks. This review deals with the most urgent issues of ecological risk. Of course, many problems require further discussion, which I hope this paper will stimulate.

References

Abrahamson, D. E. (Ed.) (1989). *The Challenge of Global Warming.* Washington, DC: Island Press.

Ames, B. N. and Gold, L. S. (1997). 'Pollution, pesticides and cancer misconceptions' in: *What Risk? Science, Politics and Public Health.* Bate, R. (Ed.) Oxford: Butterworth-Heinemann.

Boden, J. A. (Ed.) (1994). *Environmental Core. A Constructive Response to Earth in the Balance.* San Francisco: Pacific Res. Inst. for Public Policy.

Bate, R. (Ed.) (1997). *What Risk? Science, Politics and Public Health.* Oxford: Butterworth-Heinemann.

Carson, R. (1994). *Silent Spring.* New York: Houghton Mifflin Co.

Dodds, F. (Ed.) (1997). *The Way Forward. Beyond Agenda 21.* London: Earthscan Publ.

Ellsaesser, H. W. (Ed.) (1992). *Global 2000 Revisited. Mankind's Impact on Spaceship Earth.* New York: Paragon House.

Everest, D. (1997). 'How are decisions taken by governments on environmental issues?' in: *What Risk? Science, Politics and Public Health.* Bate, R. (Ed.) Oxford: Butterworth-Heinemann.

Felix, R. W. (1997). *Not by Fire but by Ice.* Bellevue, WA: Sugarhouse Publ.

Gore, A. (1993). *Earth in the Balance. Ecology and the Human Spirit.* New York: A Plume Book.

Gorshkov, V. G. (1995). *Physical and Biological Bases of Life Stability.* Moscow: VINITI Publ.

Grigoryev, A. A. and Kondratyev K. Ya. 'Global population dynamics'. *Izvestiya Russian Geograph. Soc.* **129**, 4:1–10.

Houghton, J. T. (1994). *Global Warming. The Complete Briefing.* Oxford: A Lion Book.

Inge-Vechtomov, S. G., Kondratyev K. Ya., Frolov A.K. (Eds.). (1997) *Ecological Problems of North-Western Russia and Ways to Their Resolving.* St. Petersburg: Victoria Publ..

Isachenko, A. G. (1997). Latitudinal zonality and mechanisms of landscape stability to anthropogenic impacts. *Izvestiya Russian Geograph. Soc.* **129**, 3:15-22.

Jastrow, R., Nierenberg, W., Seitz, F. *Scientific Perspectives on the Greenhouse Problem.* Ottawa, IL: The Marshall Press, Jameson Books, Inc.

Jaworowski, Z. (1997a). 'Beneficial ionising radiation' in: *What Risk? Science, Politics and Public Health*. Bate R. (Ed.) Oxford: Butterworth-Heinemann.

Jaworowski, Z. (1997b). Ice core data show no carbon dioxide increase. *21st Century Sci. and Technol.* **10**, 1:42–52.

Kealey, T. (1997). 'Should we trust science?' in: *What Risk? Science, Politics and Public Health*. Bate, R. (Ed.) Oxford: Butterworth-Heinemann.

Kondratyev, K. Ya. (1996). New tendencies in global climate studies. *Izvestiya Russian Geograph. Soc.* **128**, 6:47–54.

Kondratyev, K. Ya. (1997a). Contemporary global ecodynamics. *Studying the Earth from Space*, Issue **5**: 105–126.

Kondratyev, K. Ya. (1997b). Comments on the 'Open Letter to Ben Santer'. *Bull. Amer. Meteorol. Soc.*, **78**, 4:689-691.

Kondratyev K. Ya. (1997c). Key issues of global change at the end of the second millennium. *Newsletter European Geophys. Soc.* **63**:4-8.

Kondratyev, K. Ya. (1998a). 'Rio + 5': The Outcome of the Special Session of the UN General Assembly (23–27 June 1997). *Vestnik Russian Acad. Sci.*, **1**:30–40.

Kondratyev, K. Ya. (1998b). *Multidimensional Global Change*. Chichester: Wiley/Praxis.

Kondratyev, K. Ya., Buznikov, A. A., Pokrovsky, O. M. (1996). *Global Change and Remote Sensing*. Chichester: Wiley/Praxis.

Kondratyev, K. Ya., Donchenko, V. K., Losev, K. S. et al. (1996). *Ecology-Economics-Policy*. St. Petersburg: Russian Acad. Sci. Scientific Centre Publ.

Lavrov, S. B. Russia's geopolitical space: myths and reality. *Izvestiya Russian Geograph. Soc.* 1997, vol. **129**, issue 3, pp.1-5.

Maduro, R. A., Schauerhammer, R. (1992). *The Holes in the Ozone Scare. The Scientific Evidence that the Sky Isn't Falling*. Washington, DC: 21st Century Science Associates.

Müller, H. E. (1997). 'The risks of dioxin to the human health' in: *What Risk? Science, Politics and Public Health*. Bate, R. (Ed.) Oxford: Butterworth-Heinemann.

Nilsson, R. (1997). 'Is environmental tobacco smoke a risk factor for lung cancer?' in: *What Risk? Science, Politics and Public Health*. Bate, R. (Ed.) Oxford: Butterworth-Heinemann.

Seliverstov, Yu. P. (1997). The place of specially protected territories in concepts of sustainable development. *Izvestiya Russian Geograph. Soc.* **129**, 4:17-24.
Trofimov, A. M., Kotlyakov, V. M., Seliverstov, Yu. P. (1997). Theoretical aspects and basic principles of modelling complex ecologo-economic systems. *Izvestiya Russian Geograph. Soc.* **129**, 3: 6–14.
Van der Sluijs, J. P. (1997). *Anchoring Amid Uncertainty. On the Management of Uncertainties in Risk Assessment of Anthropogenic Climate Change.* Leiden: Ludy Feyen.
Weber, G. R. (1995). *Global Warming. The Rest of the Story*. Wiesbaden: Boettiger Verlags GmbH.
Weetman, D. F. and Munby, J. (1997). 'Benzene and leukaemia' in: *What Risk? Science, Politics and Public Health.* Bate, R. (Ed.) Oxford: Butterworth-Heinemann.
Wiin-Nielsen, A. A note on hemispheric and global temperature changes. *Atmosfera,* **10**, 3:125–136.
Wilson, J. (1997). 'Thresholds for carcinogens: a review of the relevant science and its implications for regulatory policy' in: *What Risk? Science, Politics and Public Health.* Bate, R. (Ed.) Oxford: Butterworth-Heinemann.

5 Politics, policy, poisoning and food scares

B. M. Craven
C. E. Johnson

Summary

Increasing affluence has allowed the modern reliance on outside food services for provision of food and consumption of processed rather than fresh food. Given the resultant moral hazard an increase in food poisoning was and remains inevitable. Throughout the 1990s a four-fold increase in food poisoning has been accompanied by dozens of new food regulations designed to reduce risks from food. Regulation is not achieving its desired effects in protecting consumers but is imposing significant costs upon both those whom the regulators are trying to protect; the producers and the consumers. These regulatory costs fall disproportionately on small companies. The increasing costs to the health services, the family and industry from increasing incidence of food poisoning are small by comparison with other causes of morbidity and mortality. The hugely larger costs to the nation resulting from back pain and cancer do not receive the same media treatment given to food poisoning. Because back pain and cancer are so common, they are not issues inviting media coverage.

In dealing with food poisoning a balance has to be struck between placing responsibility on the individual for their own actions and protecting the individual from the mistakes, incompetence or deceit of others; there is now too much emphasis placed on attempting to protect the individual.

Background: food scares and food poisoning

Cod liver oil, the foul-tasting vitamin supplement forced on generations of children, has been found to contain traces of dangerous industrial chemicals including some linked with

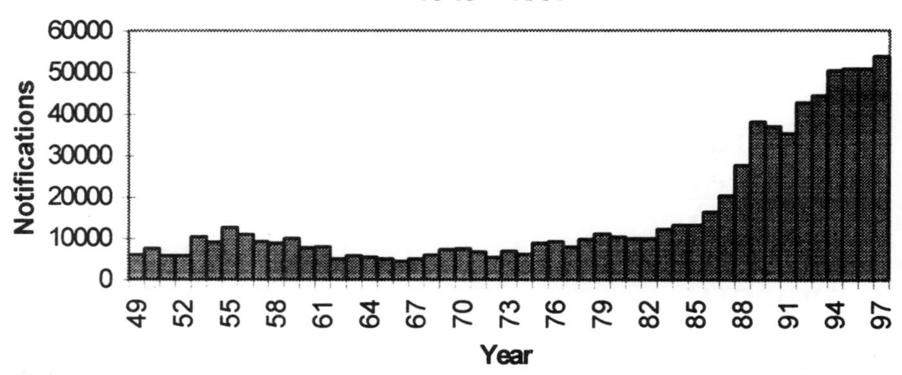

Figure 5.1 Food poisoning notifications 1949–1997 (*Source*: Communicable Diseases Surveillance Centre)

> cancer, a report said yesterday ... Sainsbury's said last night that it was considering putting a warning on bottled cod liver oil that it should not be given to children under five or adults of low body weight after discussions with the Department of Health and MAFF (*Daily Telegraph*, 20 January 1998).

Food scares arise when suddenly some food or food processes are asserted to contain a new and unexpected risk to the public health. Some of these alleged risks are treated with ridicule and contempt by the public but, others cause massive switches in consumer preferences and official policy. Sometimes the reaction is a mixture of the two. The alleged threat of contracting new variant Creutzfeldt-Jakob disease (nvCJD) after eating T-bone steaks raised the prospect of a law to ban the sale of beef on the bone. Consumers reacted by rushing to load their freezers with the product before it was removed from the butchers' shelves.

There is superficial evidence that the incidence of food poisoning is increasing. There were 53 731 notifications of food poisoning in 1997; nearly a six-fold increase between 1982 and 1997 (Figure 5.1).

Many of these will be false positives, which may be due in part to a greater awareness among doctors following circulation of the definition in 1992 (CDR 1996). The considerable increase in laboratory reports of confirmed cases of *salmonellæ* (12 322 cases in 1982 to about 32 169 cases in 1997) and *campylobacter* (about 12 000 in 1982 to over 40 000 in 1995) suggests that incidence of food

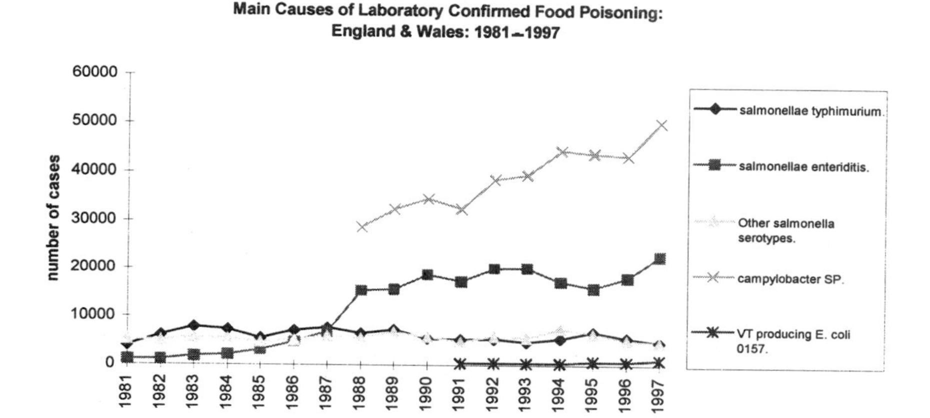

Figure 5.2 Main causes of laboratory confirmed food poisoning: England and Wales 1981–1997 (**Source**: Public Health Laboratory Service)

poisoning is increasing (Figure 5.2). The Public Health Laboratory Service (PHLS) considers that the trends are genuine because there is no evidence that clinicians have changed their criteria for submitting specimens for examination, nor that laboratories have changed in significant ways their reporting practices.

Deaths from the main forms of food poisoning rose from 187 in 1986 to 356 in 1986 (Table 5.1). Included in these figures are deaths from **listeriosis** (18 in 1996), **salmonellæ** infections (49 in 1996) and other food poisoning bacteria (1 in 1996). Deaths from **E. coli** and **campylobacter** are not reported specifically in this table, and are classified as deaths from intestinal infections due to other organisms. This category increased rapidly after the definition was introduced in 1992. Of the 357 deaths from food poisoning in 1996, 306 (86 per cent) occurred in people aged over 65 and 255 (71 per cent) in people aged over 75. Only 34 (10 per cent) of all deaths from food poisoning were in people aged between 15 and 64 years.

Spectacular examples of catastrophic poisoning, such as the Toxic Oil Syndrome in Spain, where between 1981 and 1982 about 20 000 people were poisoned and 350 died from food poisoned by the manufacturing process, are extremely rare. Newspaper headlines involving food scares have become more common during the last 15 years. Earlier scares included the 1964 incident in Aberdeen when hundreds of people suffered after eating corned beef and other meats from a local supermarket. But even when, in 1976,

Table 5.1: Deaths from main causes of food poisoning, England and Wales: 1986–1996

	1986	1987	1988	1989	1990	1991	1992	1993	1994	1995	1996
Cholera	0	0	0	0	0	0	1	0	0	0	0
Typhoid fever (**Salmonellæ**)	0	2	0	0	2	0	1	0	0	0	0
Paratyphoid fever (**Salmonellæ**)	0	0	0	0	0	0	0	0	0	0	0
Other **Salmonellæ** infections	40	52	58	61	68	62	59	35	39	36	48
Shigellosis	3	3	2	1	0	3	3	0	3	1	1
Other food poisoning (bacteria)	3	3	2	2	1	1	0	2	0	0	1
Amoebiasis and other protozoal intestinal diseases	2	1	0	2	4	2	4	2	6	5	5
Intestinal infections due to other organisms	52	26	36	54	43	34	82	79	117	172	223
Ill defined intestinal infections	76	75	65	65	69	67	90	76	57	63	61
Listeriosis	11	17	11	16	9	6	5	9	11	14	18
Total	187	179	174	201	196	175	245	203	233	291	357

Source: Office for National Statistics, Mortality Statistics, cause, 1996; 1986–1995 figures supplied from Public Sector Laboratory Service

more than 1000 people fell ill from food poisoning in Leeds in the space of three months the story failed to make national headlines (North and Gorman 1990). Two scares stand out in the UK and both proved to be disastrously costly for thousands of small agricultural producers. The first occurred when a misleading statement by a Junior Health Minister, Edwina Currie, in 1988 resulted in the slaughter of more than a million hens, mostly involving small producers, with no impact on **salmonellæ** poisoning: 'We do warn people now that most of the egg production in this country, sadly, is now infected with **salmonellæ**'. The second occurred in 1996 when the then Health Secretary, Stephen Dorrell, made a statement to Parliament officially linking bovine spongiform encephalopathy (BSE) with its human equivalent, CJD. Despite this link being only a theoretical possibility it resulted in a European Union (EU) world ban on all UK beef exports and slaughter of hundreds of thousands of the UK herd. The EU deemed that British beef was safe enough for only the British to eat.

Increasing affluence: hazards in consumption and production

The dangers from inadequate hygiene standards and food have been known about for at least 4000 years. The book of Leviticus records the laws made by Moses to protect his people against disease. These included the importance of washing hands. The eating of swine, now known to excrete **salmonellæ** and harbour **taenia solium** (tape worm), was forbidden. Other forbidden foods included mice, rats, lizards, snakes, vultures, eagles and shellfish (Hobbs and Roberts 1990).

The big advances in the prevention of ill health were made in the nineteenth century. These included measures to improve the disposal of sewage and the heat treatment of milk by pasteurisation. Pasteurisation is a common method of treating food for preservation or to destroy harmful bacteria. Food-borne infection (such as **salmonellæ** and **campylobacter**) can still occur in tuberculin tested raw milk. Chlorination of drinking water began in 1905 in Britain.

At the turn of the century poor people were ill-fed and ill-nourished. Two out of five possible recruits for the Anglo-Boer war in South Africa of 1899 to 1902, who had to be at least five feet tall, were rejected as too feeble. Whilst the government may have been indifferent to the personal health of its citizens 'there was cause to worry if national security was to depend on an army of crippled dwarfs' (Seddon 1990). By the time of the start of the First World War in 1914 the recruitment position had not changed. In the depression years of the 1930s food choice was very limited – the working class ate white bread, margarine, sugared tea, potatoes and corned beef. Those exposed to the austerity of this period knew how food keeps and decays. They knew that sugar, salt and vinegar preserve. They knew that raw meat must be treated with extreme care and that unsmoked fish had to be used within a day of purchase. When Britain entered the Second World War in 1939 two out of three called up were recruited. Malnutrition then was more often a consequence of insufficient food. Little has been heard during the past few decades about malnourishment as a consequence of lack of quantity of food. Rather, the issue has become one of nutritional food content and safety of manufactured food. Furthermore, there are other clear indications, such as height and other physical measures, overall that food quality is far better than it has ever been.

Table 5.2: Expectation of life by gender at birth, UK, 1901–1996

	1901	1931	1961	1991	1992	1996
Males	45.5	57.7	67.8	73.2	73.6	74.4
Females	49	61.6	73.6	78.7	79	79.7

Source: **Social Trends**, various issues

During the industrial age of the nineteenth century, it was common for people to witness death, particularly of the young. Stillborn deaths were much more frequent. Illness and fatal accidents were endemic and life expectancy was short (see Table 5.2).

Table 5.2 shows how life expectancy has increased steadily over the twentieth century. Increased life expectancy of this magnitude has overwhelmingly been caused by improved living standards, while the contribution of socialised medicine has been marginal. Longer life expectancy has been accompanied by improvements in general health. Increasing affluence has enabled millions of households to afford to eat protein-rich food regularly, as well as enjoy warm accommodation and good sanitation facilities. Growing affluence led the economist, J. K. Galbraith, writing perspicaciously in 1973, to speculate on future lifestyles predicting the 'contracting out of consumption from the household to the independent entrepreneur. The cooking of food by the housewife shifting to the restaurant, the food that is still consumed at home is otherwise pre-cooked or pre-prepared'.

Increased affluence has fed demand for manufactured food – from canned produce and biscuits which were novel to the Victorians, to the complete, chilled gourmet meals from supermarkets' shelves today. But mass production of food and food products bring, in turn, the need for scrupulous hygiene standards and the scope for wide scale poisoning incidents, as shown by the prominent example of factory farming of chickens and eggs where, despite regulatory controls, in the UK, **salmonellæ** is ubiquitous in poultry, as mentioned above. In another example North (1993) showed how even large slaughterhouses, like the best equipped operating theatres, are unable to eliminate cross contamination, and how regulatory attempts to do so imposed such absurdly strict requirements that meat quality fell in consequence.

Affluence also allowed the market for deep freezers to grow and mature where cooked food could be stored for long periods before use. In 1981, 49 per cent of British households possessed a deep freezer; by 1996 this had risen to 89 per cent. Affluence permitted most households to own a microwave oven which facilitated quick defrosting of pre-cooked frozen food. From no recorded statistics in 1981, by 1996, 70 per cent of British households possessed a microwave oven (**Social Trends** 1997). Food could be cooked in minutes which otherwise would have required hours of traditional oven cooking. Both these undoubtedly useful labour saving innovations, when used improperly, can provide conditions where food-borne pathogens flourish.

In addition, increasing affluence has separated the principal from the agent, as alluded to by Galbraith above. When more people eat communally in restaurants, the eater of the food, the principal, is separated from the preparer, the agent. This puts the onus on the agent to adopt the, presumably high, standards of the principal. Because of the obvious scope, and possible incentive, for the agent to adopt lower standards of cleanliness than the principal, there is a problem of moral hazard resulting in an increased risk of poisoning.

On the other hand, a higher standard of living is accompanied by higher expectations, and a greater propensity to complain if disappointed or inconvenienced. This could mean that, although the level of food poisoning incidents may remain constant, the numbers reported may not, and this may explain some of the recorded increase.

Other retailing trends also present potential hazards. Increased affluence has allowed increased car ownership. This has encouraged the growth of large out of (or edge of) town hypermarkets (Burke and Shackleton 1996). Weekly buying of food at the hypermarket has replaced frequent shopping for food at the local butcher, baker or grocer (North and Gorman 1990). Food is being kept longer in the home before consumption. Increasing affluence has stimulated a huge growth in sales of fresh chilled meals and more than a thirty per cent decline in home cooked meals since 1980. So concerned are supermarkets to avoid selling any food that subsequently poisons, they have replaced the 'sell by' date with a 'use by' date. At one Tesco store £3500 per week of fresh food is destroyed because the 'use by' date has passed. There are about 1000 stores owned by Sainsbury's, Tesco and Waitrose (**Daily Telegraph** 1998). Sainsbury's finds 2.5 tons of fresh food per week which has passed the 'use by' date, from 10 London stores, safe enough to pass on to good causes. These actions are not required by law but are the self-interested responses of a company with a good reputation to defend and promote.

Home cooked meals have also been replaced by the purchase of already pre-cooked hot food. Hot food retailing began in Britain with the fish and chip shops – still in existence in some towns. Competition came first from Chinese restaurants and later 'take aways' (to go) and, secondly, from Indian restaurants. The fast food and restaurant chain concept, such as the American burger and pizza chains, are a relatively new phenomenon in the UK. All these examples of the growth in food manufacturing and processing are potentially able to increase the number of food poisoning cases.

Given that many food hazards arise from affluence and from many sources, the objective should be not to eliminate the poisoning but to establish an environment that results in an acceptable (optimal) level of food poisoning. Too often, the Food Safety Agencies adopt policies with the target being the elimination of risk. The most ludicrous example of this is, in the UK, the illegality of the sale of beef on the bone despite the risks being microscopic. In turn, this leads not only to over-regulation, but to a distrust of the regulatory process.

By contrast, all economists, and most individuals, recognise that a balance has to be struck between placing responsibility on the individual for his own actions and protecting the individual from the mistakes, incompetence or deceit of others. In context, the problem arises from the asymmetry of information between the buyer (principal) and the seller (agent). In consumer law there is the concept of **caveat emptor**; let the buyer beware. In these examples the information known by the seller is much more comprehensive than that known by the buyer and there is an incentive for the vendor to withhold information. However, in practice, both the consumer and the producer of food rely on reputations. These issues are addressed in subsequent sections.

Fear and death

Public concerns over food safety fall predominantly into three categories:

1. Contamination by infectious agents

2. Contamination by chemicals
3. Nutritional concerns

Nutritional concerns include fat, salt, sugar, cholesterol, and junk food (with concomitant overeating/obesity and poor nutrition, which are generally recognised as major contributors to degenerative diseases such as heart disease and diabetes). However, the public does not exhibit a great deal of fear regarding these issues and they do not tend to escalate into major food scares. By contrast, when viewing Tables 5.1 and 5.3, which list the numbers of deaths from selected causes, it is apparent that deaths ascribed to contamination by infectious agents account for a minuscule number of the total, and deaths from contamination by chemicals are not recorded at all. Yet the public experiences a great deal of fear and apprehension over possible microbial and chemical contaminants and many disastrous food scares have resulted.

The fact is that food poisoning threatens only the elderly and young and those with compromised immune systems. There are more deaths every year from people drowning in their bathtub than of food poisoning in those aged between 15 and 64. As with AIDS, the general public is not at risk.

While lethality of the risk is an important factor in how the public assesses its severity, it is not the actual lethality that counts, but rather the perceived lethality. Deaths from nuclear power plants have been very few, and the actual risk very small indeed (Table 5.4), but perceived risk of death is high because the public thinks in terms of the potential for huge numbers of deaths if something goes wrong.

Another example is the food scare involving BSE, a disease that is tenuously linked to nvCJD, an extremely rare condition that has accounted for only 22 deaths in total (March 1998) in the UK. The fatality of the disease, combined with media speculation about impending huge epidemics of nvCJD, led to a very high perceived risk among the public. Fear of nvCJD was also no doubt influenced by the tendency of people to over-estimate deaths from rare diseases while deaths from common diseases are underestimated (Slovic et al. 1981).

Experts versus the public

It has been well established that experts differ tremendously from the public in the way they view and assess risk (Slovic et al. 1981,

Table 5.3: Death by selected causes (England and Wales) 1996

Cause		
All Causes		560 135
Malignant neoplasms		139 459
Digestive organs and peritoneum	38 226	
Trachea, bronchus and lung	32 273	
Breast	12 179	
Ischaemic heart disease		129 047
Cerebrovascular disease		55 021
Pneumonia		54 137
Diabetes		5994
Suicide		3445
Motor vehicle accidents		3134
All other accidents	8368	
Accidental falls		3637
of which:		
Fall on or from stairs in home	558	
Slipping, tripping or stumbling	144	
Fall from chair or bed	83	
Fall into hole in ground	4	
Accidental poisoning		1089
of which:		
by medical drugs, medicaments and biologicals	937	
by alcohol	152	
by utility gas in the home	70	
Choking on food		262
Meningococcal infection (meningitis)		245
Accidental drowning		216
Struck accidentally by falling object		47
Clothes catching fire		45
Choking on objects other than food		43
In the bathtub		35
All food poisoning other than *E. coli, listeriosis* and *salmonellæ* (persons aged 15–64)		**22**
Injury caused by animals		12
Plastic bag		9
***Listeriosis* (persons aged 15–64)**		**6**
***Salmonellæ* (persons aged 15–64)**		**6**
Hornets, wasps and bees		5
Other venomous arthropods	1	
In sports		5
Lightning		4
Dog bite		3
Fireworks		3
E. coli (persons aged 15–64)		**0**

Source: Office for National Statistics

Table 5.4: Risk of an individual dying in any one year from various causes

Smoking 10 cigarettes per day	1 in	200
All natural causes age 40	1 in	850
Any kind of violence or poisoning	1 in	3300
Influenza	1 in	5000
Accident on the road	1 in	8000
Accident at home	1 in	26 000
Accident at work	1 in	43 500
Radiation working in radiation industry	1 in	57 000
Homicide	1 in	100 000
Poisoning from **salmonellæ** in poultry meat	1 in	5 000 000
Hit by lightning	1 in	10 000 000
Release of radiation from nuclear power station	1 in	10 000 000
Eating beef on the bone	1 in	1 000 000 000

Source: BMA (1987), Maitland (1998)

Fife-Schaw and Rowe 1996). For example, on a list of 30 hazards, experts rated nuclear power generation at the lower end of the scale (less risky), believing it to be approximately as dangerous as food colouring and home appliances, and less dangerous than food preservatives, riding bicycles, and swimming. On the other hand, the public rated it at the number one position (most dangerous of all) (Slovic et al. 1981). Experts are least uncomfortable with a hazard that has the potential to kill many people, but which has a very low probability of happening (such as massive nuclear power radiation or meltdown), whereas the public feels more comfortable with a hazard that affects fewer people, such as a road accident, but which has a much higher likelihood of occurring. This accounts for the fact that what experts communicate about the low risks involved in various food scares is often almost totally disregarded by the public. An example of this dynamic is a major food scare in the 1980s which involved daminozide (Alar), a ripening agent used on apples. Experts considered this agent to be harmless, but the safety of its breakdown products was under investigation (Lee 1989). The Natural Resources Defense Council, a consumer advocacy group, announced that Alar in apples created a 45 in a million cancer risk, and predicted that 6000 American school children could get cancer from Alar residues. This created a media stampede and public hysteria, and apples were banned from school lunchrooms across the nation and removed from supermarket shelves. Eventually, the dust cleared after the increased cancer risk from Alar had been calculated at 0.025 per cent (considered to be trivial) and apples were restored to school menus. However, 'No danger from Alar in apples', doesn't make a newsworthy headline, so damage from media coverage of the scare was not reversed, but left a lingering psychological impact. Ultimately, trust in foods was damaged and losses to the apple industry were estimated at over $100 million (Lee 1989).

The outrage component

Experts may often feel frustrated that the public ignores the facts and bases its opinions on what seem like illogical thought processes. However, the public's method of risk assessment involves many complex cognitive processes that need to be understood and addressed in order to communicate risk in a way that people will understand and accept. Sandman (1994), America's foremost expert on risk communication, has developed the following formula to explain this phenomenon:

RISK = HAZARD + OUTRAGE.

According to Sandman, traditional risk assessment defines risk as magnitude (how bad the problem could be) times probability (how likely it is to happen): 'However, experts tend to focus on this definition, (let's call it hazard), and so underestimate actual risk, because they ignore outrage. The public tends to focus instead on outrage and pay less attention to hazard' (Sandman 1994). Most public policy disputes over food safety are concerned with whether a given risk is **acceptable**. That value judgement depends on outrage factors more than on the size of the risk (Groth 1991). In order to communicate effectively a risk to the public, it is necessary to focus on the outrage. Sandman has identified 12 components of outrage and how they affect perceived risk.

Once again, using nuclear power plants as an example, the outrage components fall almost exclusively into the right column, creating a high degree of outrage, and concomitantly a high level of fear and a low level of acceptance. Food hazards involving chemicals and microbial contamination also score high on the outrage scale. Other analysts (Slovic et al. 1981, Fife-Schaw and Rowe 1996) have pinpointed the three most important categories as being: first, dread, secondly, the unknown and thirdly, the number of people exposed.

Table 5.5: Public risk perception

Less perceived risk		**More perceived risk**
Voluntary	vs.	Coerced
Natural	vs.	Industrial
Familiar	vs.	Not familiar
Not memorable	vs.	Memorable
Not dreaded	vs.	Dreaded
Chronic	vs.	Catastrophic
Knowable	vs.	Unknowable
Individually controlled	vs.	Controlled by others
Fair	vs.	Unfair
Morally irrelevant	vs.	Morally relevant
Trustworthy sources	vs.	Untrustworthy sources
Responsive process	vs.	Unresponsive process

Source: Communicable Disease Surveillance Centre

The 'dread' component is of particular importance. Perceived risk can be predicted almost perfectly from first, how much a hazard is dreaded and secondly, how lethal the hazard is considered to be (Slovic et al. 1981). 'Dread' includes such things as greater public concern, more serious effects on future generations, threats of widespread disastrous consequences, and potential to become more serious (Sparks and Shepherd 1994). According to Slovic, the higher the dread rating of an activity, 'the higher its perceived risk, the more people want its risk reduced, and the more they want to see strict regulation employed to achieve the desired reduction in risk'.

All of the above explains why nutritional concerns don't generate widespread public fear and huge food scares, whereas microbial and chemical contaminants do. Food hazards such as alcohol, sugar, fat, nutritional deficiencies and so forth adversely affect large numbers of people, yet they rate very low on the 'dread' scale, and are so very familiar that they rate very low on the 'unknown' scale as well (Sparks and Shepherd 1994). Conversely, chemical contamination rates very high on all three scales (making it no surprise that Alar in apples generated such a massive food scare). Contamination by infectious agents rates high on the dread scale but low on the unknown scale (Sparks and Shepherd 1994). No matter what the actual risk, hazards which rate high in outrage are most likely to become scares.

If it bleeds, it leads

What explains media attraction to food scares? How does a concern over food safety escalate into a food scare? Obviously, the media often play a crucial role in this dynamic. Which issues the press decides to cover, how it covers them, and how much coverage it gives them, often determines whether the issue becomes a scare. Is it the seriousness of a food hazard (the potential risk) that attracts the media? Not necessarily. Sharon Begley, science editor at **Newsweek** magazine, states that 'Risks surrounded by uncertainty and controversy are good contenders for coverage' (Begley 1991). Traditional journalistic criteria, such as timeliness, proximity, prominence, human interest, drama and visual appeal, make a big controversy intrinsically newsworthy even if it's not a serious health threat (Sandman 1994).

When asked why journalists write about some risks and not about others, Begley stated that 'Journalists are paid to engage their audience ... We can't do it by boring them. One way to avoid boring people is with stories that have a strong element of uncertainty. Reporters crave uncertainty because it is generally equated with drama ... Most of what science "knows" about risks such as pesticides or hormones in food or food irradiation is uncertain and highly debatable.' Begley also says that journalists 'seek out the contrarian', that they are attracted to 'subjects, discussions, and controversies that go against conventional wisdom and that will evoke in our readers a sense of surprise.' Unsuspected risks in food make good stories because they oppose the conventional wisdom (that food in first world countries is safe). It's really not news that smoking still causes cancer or that driving without seat belts is still a foolish thing to do, but new food hazards always have potential for a good story.

The media play into what Slovic (1981) calls 'availability bias'. This means that people will judge an event as more likely to occur (and thus be more apprehensive) if it is easy to imagine or recall. Even though frequently occurring events are easier to imagine or recall, the availability bias will often come into play regarding events that don't occur frequently, or perhaps don't occur at all. As Slovic points out, 'Discussing a low-probability hazard may increase its imaginability and hence its perceived riskiness, regardless of what the evidence indicates' (Slovic 1981). Availability is similar to

Sandman's item on the outrage scale 'memorable vs. not memorable'. Since the media tend to cover events that are more dramatic (even though the actual risk may be low), and since they cover these events more often, media coverage increases both availability and memorability. In an attempt to ascertain how much the media influences risk perception, two groups of people were asked to rate items on a list of some 90 hazards according to how risky they believed each hazard to be (Kone and Mullet 1994). One group consisted of Burkina Faso intellectuals, the other of French students. France and Burkina Faso are markedly different in terms of geography, economics, politics, and ethnic background, but quite similar as to media coverage. There are also extreme differences in the real risks that exist in either country (many risks one might be exposed to in third world countries do not occur in first world countries). Even so, the media in Burkina Faso give coverage to hazards that may not even be present there (but would be of concern to citizens of France). The results of the study were that the Burkina Faso inhabitants had 'approximately the same preoccupations as the French respondents and to the same degree'. This indicates that reality plays a secondary role to the media's representation of reality.

But the decisive factor is that food hazards involving chemicals and microbial contamination inspire outrage. 'Outrage is newsier than mere risk. Outrage is active, risk is a number that just sits there' Begley (1991). 'The media are in the outrage business. Most of the coverage isn't about the risk. It's about blame, fear, anger, and other non-technical issues-about outrage, not hazard ... In their focus on outrage rather than hazard, journalists are at one with their audience' (Sandman 1994).

Consider the source

Sandman did a content analysis study on 248 environmental risk stories from New Jersey's 26 daily newspapers (Sandman 1994) and found that 68 per cent of the paragraphs contained no risk information at all, another 15 per cent dealt with whether the potentially risky substance was present or absent, and only 17 per cent of the paragraphs dealt with whether the substance was risky or not. A panel consisting of an environmental reporter, an activist, an industry spokesperson, and a technical expert was asked to assess these stories further. Although members of the panel disagreed about almost everything else, they strongly agreed that 'environmental risk information was scanty in these stories. Technical content was especially lacking. What risk information was provided came mostly in the form of opinions, not evidence'. Government is by far the most common source of environmental risk information used by the media. In the New Jersey study, government officials accounted for 57 per cent of all source attributions. On network television, government represented 29 per cent of the on-air sources, but when only one source was used, it rose to 72 per cent (Sandman 1994). According to Sandman, 'Reporters typically start with a government official, the swing vote. If the government says "dangerous", they look for an industry source or possibly an expert to say "safe". If the government says "safe", they look for a citizen or possibly an activist to say "dangerous".' In view of the above, it is of particular interest to note that government officials are the source which is least trusted by the public. On the hazards of both food poisoning and excessive alcohol use, a committee of medical doctors was rated as highly credible, whereas government was rated as having low credibility (Frewer 1997). Rating for distrust, members of parliament, government ministers, and government ministries were rated among the top four most distrusted sources, along with tabloid newspapers (Frewer 1996). It is not surprising to see how food scares escalate when most information about them comes from the least trusted sources.

BSE: a case history

Considering the profound abyss between the actual risk of contracting nvCJD from eating beef and the public's perception of this risk (with government regulatory response almost totally divorced from reality), the BSE food scare is clearly the worst in history. In 1994, BSE rated high on the 'dread' scale but only in the middle of the 'unknown' scale, whereas other microbial contaminants, such as **salmonella**, rated low on the unknown scale (Sparks and Shepherd 1994). However, in 1995 the picture changed dramatically when ten cases of nvCJD were identified among the 207 cases of regular CJD neuropathologically examined since May 1990 (Morabia and Porta 1998).

In 1996, the alleged link between BSE and nvCJD was given official sanction. The food scare escalated, with concomitant public pressure on the government to regulate the risk. In response, regulatory action accelerated into the realm of senseless overreaction. The 'unknown' factors increased substantially: Little was known about this new variant of CJD; nothing was known about whether there would be an epidemic of nvCJD and if so, how big it would be; and finally, scientific consensus as to any real causality between BSE and nvCJD was lacking. This left the public afraid and confused (high dread, high uncertainty), the worst situation possible in terms of avoiding a food scare centred round British beef.

How long the public remained scared is another story. Although sales of beef had initially plummeted, they eventually returned almost to normal, and by the time the ban against beef on the bone was initiated, consumers were lining up at the shops to stock up before the ban took effect. By this time the scare had taken on a life of its own and the government, in order to avoid blame in the extremely unlikely eventuality of a bona fide epidemic of CJD, continued to over-react. Continued reliance on experts who offered hypotheses, not definitive knowledge, and who have previously proven dramatically wrong in their predictions of catastrophic epidemics (AIDS, swine flu, Legionnaire's disease) led the **Daily Telegraph** (27 March 1996) to opine: '... any politician who attempts to rely on [experts'] advice as a substitute for his own judgement, let alone common sense, is liable to come badly unstuck.'

Both BSE and CJD (as well as kuru and scrapie in sheep) belong to a tiny group of diseases, the spongiform encephalopathies, which are characterised, in contrast to traditional infectious diseases, by their rarity, restriction to certain locations, and inexplicable causes. Of about 160 000 recorded cases of BSE worldwide, 99 per cent are in Britain (Craven and Stewart 1997). Kuru was discovered among the Fore Indians of New Guinea, has never been documented anywhere else in the world, and eventually disappeared in New Guinea as well. Both kuru and BSE, and now nvCJD involve the ingestion of flesh. The Fore Indians allegedly practised ritual cannibalism, and many cattle herds had been fed bovine offal from BSE-afflicted cattle, which supposedly transmitted an infectious agent.

These conditions are prime candidates for being blamed on unconventional agents: at first slow viruses (lentiviruses) were blamed, then the equally hypothetical prion, an unusual protein form, came into favour as the aetiological agent. However, when years or decades must elapse between cause (eating British beef) and effect (nvCJD) it is almost impossible to prove causation, since any number of important confounding variables (non-infectious causes) may be introduced during the long latency period.

In any event, this theory was espoused by several prominent scientists, who predicted a massive epidemic of nvCJD. This, as well as the EU ban contradicted the prevailing opinion that British beef was safe (Craven and Stewart 1997). At this point, the scare escalated dramatically. Oddly enough, British citizens were still allowed to eat the beef that had been deemed unsafe for others. It would have made more sense to extend the ban to everyone and let the British eat only imported beef.

Even though a tenuous link between BSE in cows and nvCJD in humans exists, epidemiology does not support a common infectious cause. The epidemiological link is weak because it is based only on the temporal and geographical association between two diseases in whole populations (Skegg 1997). Proponents of the BSE/nvCJD hypothesis would want personnel at highest risk of exposure (those working in abattoirs, veterinary practice, farms, meat markets, butcher's shops, and laboratories) to show an excess of nvCJD, yet this is not the case. A case of nvCJD was recently reported in the UK of the death of a woman who had been a vegetarian for over 12 years. She had stopped eating meat one year before the first clinical case of BSE was reported in cattle. Of course, the concept of long latency periods could allow beef as a causal agent, even after 50 years of vegetarianism!

The handful of currently recognised nvCJD cases was documented by extraordinary effort and a worldwide search. The BSE/nvCJD question particularly suffers from several problems common to epidemiological studies: too few subjects for confident conclusions, failure to control for important confounding variables, no data (or little data) on exposure, inadequate diagnosis, and subjects who are qualitatively different from the population to be protected (Lave 1987, Skegg 1997). Scientific work to date confirms that the origin and sudden spread of BSE in cattle herds are probably explained by the introduction of forced and stressful feeding of animal protein to obligatory herbivores. Common to all forms of spongiform encephalopathies, prions have been found in the brain; however, it is possible that these prions arise as a result of the disease, rather than causing it (Axelrad 1998). In any event, spongiform encephalopathies are so rare in humans as to make then clinically irrelevant to the general populace.

Morabia and Porta (1998) stated, 'Today, the 22 cases of nvCJD more closely resemble a cluster of a rare disease than the prodromal phase of a large epidemic.' They asserted the wisdom of scientists admitting their ignorance on the subject, since 'crying wolf' in the event of the likely non-epidemic would undermine credibility on other subjects where more certainty existed. Only in October 1997 were any papers published that claimed to furnish clear-cut evidence that nvCJD was an infectious disease caused by the same agent as BSE (assuming that BSE is an infectious disease). The respective authors admitted that this was still a hypothesis, that 'considerable controversy remains', (Hill et al. 1997) and that their study shows 'compelling evidence of a link', without claims of definitive proof (Bruce 1997). Even though this work may show a common biochemical factor linking BSE and nvCJD, it does not show that they are the same disease or that either disease is caused by an infectious agent directly transmissible across the species barrier between cattle and humans. Neither does it exclude the possibility that prions and other signs in animals or persons with spongiform encephalopathies might be due to an entirely different, non-infectious process. This severe food scare ran almost its entire course without any scientific proof that there was anything to be scared of. It is reasonable to conclude that after 1996 the outrage component of public risk perception of BSE/nvCJD would have increased substantially. This, of course, attracted more media attention, which in turn increased public apprehension, which in turn increased pressure on government to 'do something'. This something took the form of grossly exaggerated regulatory response.

Food law and the regulatory framework

Principal Acts of Parliament:

The Food and Drugs Act 1938.
The Defence (Sale of Food) Regulations 1943.

The Food and Drugs Act 1955.
The Food Act 1984.
Food Safety Act 1990.

In an excellent summary Thompson (1996) described the regulatory changes which began in 1938. It was then that food poisoning became notifiable under the Food and Drugs Act 1938. This Act enabled ministers to implement regulations to control the composition and labelling of foods. It also enabled Local Authorities to make bye-laws on the sanitary conditions of food for human consumption. The food shortages resulting from the Second World War gave rise to food substitutes; chicory in coffee for example, and developing technology enabled the possibility that substitutes could be passed as the real thing. The Defence (Sale of Food) Regulations 1943 strengthened the law on labelling and food content. The Food and Drugs Act 1955, had two main objectives: to safeguard health and prevent deception and fraud. By this Act, Ministers were empowered to make regulations 'in the interests of public health or otherwise for the protection of public health' (Thompson 1996). As we have seen, by the 1980s, food manufacturing became more complex, sophisticated and technological. Economic growth had generated markets for pre-packed and pre-cooked foods, the deep freezer and the microwave oven (see earlier section) and it was felt that legislation required updating. As a consequence the Food Act 1984, which replaced the Food and Drugs Act 1955, was introduced to deal with these developments. The objective of the Food Act 1984 was to ensure consumers receive a pure and wholesome food supply. The Act prohibits food from being sold which:

1. Has been made injurious by addition or abstraction of substances or by treatment.
2. Is not of the nature, substance or quality of food of a different kind, or adulterated or inferior.
3. Has false or misleading labelling or advertisement.
4. Is unfit (decomposed or contaminated).

Enforcement allowed premises to be closed within three days and remain closed pending the outcome of legal proceedings under the Act. Despite this Act, incidents of food poisoning continued to increase in the 1980s (Figure 5.3). The Government believed these

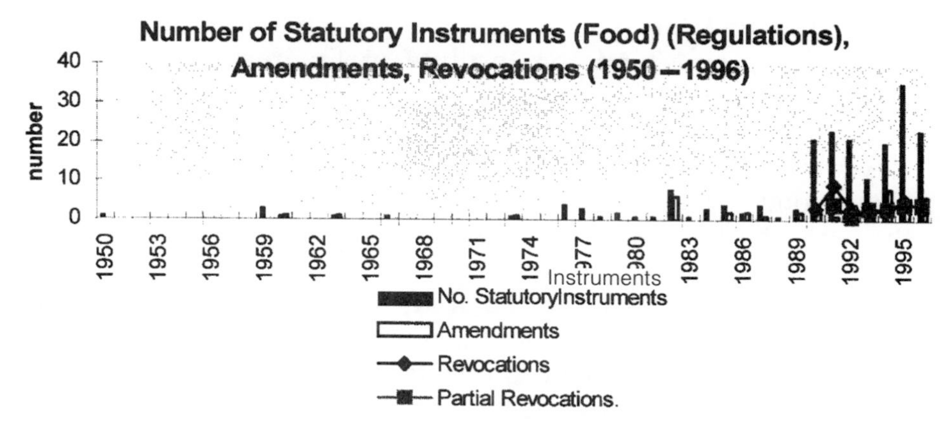

Figure 5.3 Number of statutory instruments (Food) (Regulations), amendments and revocations 1950–1996 (*Source*: Halsbury's Statutory Instruments (1996))

incidents were preventable and, a White Paper, *Food Safety – Protecting the Consumer*, (CM 732) was published in 1989. The White Paper had several main objectives. These included ensuring that new methods of production and distribution were safe. A second objective was to prevent food being mislabelled or misleadingly presented. A third objective was to ensure that European Community directives could be implemented in UK law, and a fourth to set standards that spelt out consumer rights. Other changes in the law were proposed too. These included reinforcing penalties against law-breakers and measures to strengthen controls to prevent unfit food reaching the consumer. Responsibility for enforcement was given to Local Authorities (Scott 1990) with power to charge for inspections.

Despite the objective of increasing the regulatory powers the 'White Paper portrayed a legislative system which was working extremely well' (Thompson 1996). Scott (1990) stated that the promotion of the new Food Safety Act based on the findings of the 1989 White Paper was little more than political expediency. The new Act reasserted the old law with some changes. The most conspicuous change was the addition of the word 'safety' to the title. Another change was additional funding (£30m) for enforcement. The third main change was the establishment of a Food Safety Directorate under the Minister for Food. Although this was intended to separate the interests of the food industry from the safety interests of the consumer, the Directorate remained a division of the Ministry of Agriculture, Fisheries and Food (MAFF). Consumer groups and the Institute of Environmental Health Officers were dismissive of this change. Scott (1990) stated that the Act 'provide[d] an effortless way to convey the impression that the Government [was] taking serious measures to address concerns about food safety'. The Act received uncontroversial passage through Parliament, with only irradiation of food being a contentious issue.

Although the 1980s showed a rise in the number of food regulations by comparison with the previous 30 years, this rise was small by comparison with the unprecedented increase in the number of food regulations following the Food Safety Act (1990). It is not in doubt that there are substantial costs to the tax-payer and industry of attempting to regulate against food poisoning illness. There are also costs to the health sector from food poisoning. General practitioners treat cases in the community, hospitals treat at accident and emergency units and there are follow up costs. Whether the regulatory approach that does not tackle the main cause of the increase in food poisoning (and largely preventable by adopting good hygiene standards in the handling and heating of food) is cost effective, is a moot point. It is of little surprise that the stronger regulatory regime has not been accompanied by a fall in food poisoning. In short, the UK is awash with regulations designed to improve food safety. All the requirements of these regulations are sensible, but do they require such immense legal status and force?

Theory and politics of regulation

Food poisoning imposes costs on employers, the health services, the family and society in taking care of those who become ill. It was estimated that 10 000 working days were lost in the survey reported by Roberts and Sockett (1994). Whilst this appears to be a huge cost, it pales into insignificance by comparison with the 116 000 000 working days lost through back pain (NPBA). Roberts and Sockett estimated the production loss to industry from sickness related absence from work to be a maximum of £334 million. By comparison, the estimated cost to industry from back pain was £5100 million (NPBA). The response to the consequences of food poisoning, as we have seen, has been a deluge of regulatory statutes. In context regulation is perceived as necessary for requiring or controlling a process or treatment used in the preparation of food intended for human consumption (Schefferle 1990). Regulation may also be intended to generate consumer information (labelling, contents, processes used). It is rarely denied that regulation is other than well intentioned in democratic countries and designed to protect public and private interests (Tower 1993, Veljanovski 1991). The reality is often quite different from these good intentions.

Market failure

Food scares and consumer problems with food have been shown to be based largely on a mixture of the following elements: consumer ignorance about the risks involved, lack of information on product content, misleading information on product content. These are elements of 'market failure' where the consumer has not made an 'optimal' decision due to lack of full information. It is an important issue in that governments seek to make up for this failure by ensuring that the necessary information is provided. In order to achieve this, a government must determine whether the greatest food problems (and hence potentially greatest scope for successful problem solving) are associated with a lack of information, or are caused by inadequate hygiene standards, 'bad' food products or even, perhaps, the consequences of regulatory action?

There are two main views of how to correct 'market failure.' The first (and minority view) is that in any dynamic nature of society, where technology is changing and information is always very imperfect, a minimally regulated competitive process produces a result better than any other approach. The alternative, and still dominant, orthodoxy differs in that it advocates government intervention, through state supply, fiscal measures or regulation, to correct the failures of the market. It is generally understood in a theoretical sense that regulatory interventions must not be so severe as to impose costs which outweigh benefits or which would drive businesses abroad. In practice, cost-benefit approaches to food safety regulation are seldom adopted or even attempted. It appears that regulators often appear to fail to account for costs arising from regulation, and on whom these costs fall. These costs can be so great that they make the situation worse ('public failure'). The costs arise not only from the regulatory authority in its day-to-day business of regulating, but also in compliance costs imposed on the regulated (Veljanovski 1991). A good example occurred when the British Agricultural Minister, after stating on 3 December 1997 that 'our beef is the subject of the most rigorous safeguards of any beef, anywhere in the world', announced that the sale of beef on the bone was to be banned. The estimated cost of removing the bone would push the price for prime sirloin up from £4.50 to £6.50 per pound. This was added to an estimated £70 per animal in compliance costs of existing government regulations on beef.

Principal versus agent and asymmetric information

Regulation is predicated on the belief that experts or the government have superior knowledge to the consumer. In practice, politicians, when making decisions on complex issues have to rely on a public sector bureaucrat for a supply of filtered information. The bureaucrat in turn has to acquire summarised and filtered information from experts. In attempting to act to improve the public health, Government decision-makers face 'bounded rationality' (Williamson 1975) whereby people are generally severely limited in their ability to process all relevant information affecting a complex issue. Decision-makers are therefore unlikely to be aware of the subtleties associated with the regulatory functions they are recommending and funding. The process of regulation therefore is one where Governments request and receive expert advice and often, after consultation with civil servants, make policy decisions. The process assumes that experts are neutral and objective and that experts and policy makers act rationally. Governments use the process too to legitimise their actions and thus can avoid blame when errors and failures occur.

The public and the private interest can, unfortunately, easily conflict especially where there is market power in the form of information asymmetries. A good example of such asymmetry can be found in health care markets where the producer (physician) invariably has much greater information than the consumer (patient). Other examples include the markets for financial insurance services, electrical and motor vehicle repairs. In such cases there is a suspicion, by the consumer, of exploitation by the producer. This suspicion enhances calls for 'something to be done'.

More pertinent in the context of this chapter is the likelihood of consumer ignorance and the lack of any information in the face of an alleged risk from consumption of a particular food product. When such alleged risks arise (and are given tabloid coverage) a government spokesman is obliged to give reassurance. Information asymmetries between government departments and their expert advisers generate suspicions in the public's mind of hidden agendas, especially given that politicians are some of the least trusted people in society. This arises when there is doubt about the probity of government advice and information where government departments represent conflicting interests. This was a real problem in the UK following the BSE scare in 1996 when it was recognised that the government agriculture ministry, MAFF, despite reassurances to the contrary, in effect, served both to further producer interests and protect the consumer. In 1998 a White Paper, *The Food Standards Agency: a force for change*, proposed an independent agency (although compulsorily funded by the food industry) with sweeping powers to secure food safety. Another problem is that experts may not be reliable advisers when problems occur which are new, unpredictable and either unprojectable, such as with nvCJD, or projectable with wide confidence limits, as with AIDS in the early stages of the phenomenon. Indeed in the case of AIDS, reliance of governments on imprecise information from medical experts was the main explanation for serious resource misallocation and misuse (Craven et al. 1994, Craven and Stewart 1997). The erroneous view of many experts that most cases of *salmonellæ* poisoning were caused by *salmonellæ* infected eggs is another example. Experts have, by definition, confined their knowledge to a specific topic. That is precisely why they are not qualified to assess the importance of their own topic as part of a larger scheme.

Regulatory capture

Whilst regulation is intended to protect the public from exploitation by suppliers with market power, it can also be used to protect or enhance the interests of the regulated. For example, professional qualification is a form of public regulation designed to impose minimum standards for treatment, but is also a form of private regulation because it enhances the economic rents (incomes) of the regulated, for example, the health care staff (Stigler 1971). This can lead to what is known as regulatory 'capture' where the regulated may eventually control or 'capture' the regulators (Kay 1988). A survey by Cannon (cited in Scott 1990) found that of 370 people sitting on Government advisory committees involving food, only a hundred advisors were independent in having no links to the Government or the food industry. Of the rest, 133 worked, or had worked, for the food industry; 65 were funded by, or were advisors to, the food industry and 156 had links to the British Nutrition Foundation funded by the industry. These circumstances are propitious for regulatory 'capture'.

Empire building

In practice, the efficacy of regulation is likely to be compromised for several additional reasons (Posner 1974). Regulators may fail to use their power over the regulated, or may use it in unnecessary or harmful ways. In the field of public health, North (1992) demonstrated how legislation imposed unnecessary costs on the regulated. Regulators, if beneficiaries of political patronage (*Financial Times* 1994), may not be the most appropriate people to appoint. More important is the recognition that regulators themselves cannot be presumed to be without self-interest. They have an interest in maintaining their jobs, or enlarging their budget by taking advantage of their ability to disguise costs or their inability to cost activities. This will result in the service being supplied at greater than minimum (productive efficiency) cost (Niskanen 1971, 1973). There is also an incentive for regulators to overstate the importance of their work and to ensure they do not remove the need for regulation. If the problem they are attempting to control worsens, there is a call for greater funding. If the problem diminishes, it is claimed as a success of the regulation; funding must continue. The response of the Department of Health (DoH) to the failure of AIDS to spread to the heterosexual population is a first class example. There is strong evidence that individuals are not adopting safe sexual practices. There has been a continued increase in visits

(547 437 in 1984, 732 000 in 1994 in England and Wales) to clinics for treatment of sexually transmitted diseases and constant figures of 150 000 abortions per year in England and Wales. In 1996 there were 8000 pregnancies to girls under the age of 16, the highest for 10 years. Despite this, the DoH consistently asserts that health education expenditures have succeeded in persuading people to adopt safe sexual practices. The DoH then further asserts that this is the reason why AIDS has not spread heterosexually.

Nonetheless, it is unwise to assume a homogenous set of utility maximising bureaucrats (Downs 1957, 1967). Some, for example, will seek to conserve power, prestige and budget. Others will seek to maximise these attributes to gain promotion. Some will combine loyalty, altruism and self interest. Altruistic bureaucrats may seek to act in the public interest perhaps out of loyalty to the state (the classical civil servant), subject to conserving power and influence. Taken together, it is certain that the cost of activities supplied by the bureaucrat will be excessive. Bureaucrats have much managerial discretion over what activities they choose to perform because they need not respond directly to the customer or taxpayer. Sometimes this will result in satisfying behaviour where the tasks performed suit the bureaucrat rather than the public health. At other times they will result in empire building. The public caricature of the lazy bureaucrat is misplaced. Staff will often be employed to work long hours, at a fast pace perhaps doing a particular job well, even though that may mean searching for lost memos. The essential point is that often activities will be performed which are of little or no value to the customer or taxpayer. Given the additional problem of output measurement, annual budget allocations are likely to be based on incrementalism, where the budget increases by a small amount each financial year rather than on the size, nature and seriousness of the problem to be addressed (Wildavsky 1973, 1975). In all cases, there is a supply of regulatory activity that is excessive and inefficient in both accounting and economic terms.

In recent years a new type of bureau, the SEFRA (Self Financing Regulatory Agency), and bureaucrat has emerged. The name was coined by Booker and North (1994). The SEFRA is an agency which can charge those it is regulating for the costs of the regulation. Examples include Her Majesty's Inspector of Pollution, the National Rivers Authority, the Medicines Control Agency, and the Veterinary

Medicines Directorate. The SEFRA arises from an ideological belief that regulatory agencies must be self-financing. Recall that the need for regulation arises from the demands of the consumer for protection from exploitation from information asymmetries in the market. It must follow therefore that the consumer is willing to pay, in higher taxes or higher prices, for the costs of regulation. In giving SEFRA officials the authority to charge producers for their services and the power to close producers unable or unwilling to pay or comply, a powerful force for generating rents from unnecessarily restrictive regulatory activity results. The current fees, for example, authorised under the Dairy Products (Hygiene) (Charges) Regulations 1995 SI 1995/1122 are £94 for a general dairy farm visit and £63 for a sampling visit. The claim that the costs for inspection of licensed slaughterhouses required under European Union Directive 93/119/EC are unduly onerous and possibly illegal (Thompson 1996) was tested in *Woodspring District Council v Bakers of Nailsea Ltd* (unreported). The charging was found not to be illegal but the judge did consider there were arguable grounds for challenging the Directive and Regulations which required some inspections to be carried out by veterinarians rather than by cheaper meat inspectors. When severity of regulatory controls differs between countries, the producers in countries with strongly enforced regulations will be at a competitive disadvantage by comparison with producers in countries with liberal regimes. There is evidence that this is becoming a real problem in England and Wales. Booker and North (1994) have documented dozens of cases where European Union Directives have been translated into UK law with much more onerous regulatory force than required by the Directives.

Defusing a food scare

Given the goals of the press, it is not surprising that alarming content about risk is more common than reassuring or intermediate content (Sandman 1994). However, it is sometimes difficult to determine what will be construed as alarming or reassuring. Industry tends to construe mildly alarming stories as being highly alarming. Experts consider test sample results showing low levels of contamination to be reassuring; however, many lay persons focus more on the presence of the contaminant than on its concentration and find the same data exceedingly alarming (Sandman 1994). Sandman reports that 'Explicit statements by official sources minimising the risk ("the levels are low", "it hasn't spread", "don't worry") are often considered offensive, incredible and therefore alarming by citizen readers. One-sidedly reassuring risk information is likely to strike readers as incredible and therefore produce a boomerang effect'. Once a scare has emerged, reassurances from government and scientists that the risk is actually quite small will be ineffective in subduing panic and de-escalating the scare. The outrage must be dealt with in an effective manner.

The science of risk communication is still relatively new, though valid and effective precepts have been clearly defined. Mitigating a hazard itself does not mitigate the outrage about the hazard. To defuse a scare, outrage must be addressed, that is, the public's particular concerns must be addressed and dealt with in a way responsive to their emotional needs regarding the issue. Sandman, who serves as a consultant to industry, gives the following example of good risk communication:

> About six months after the Exxon Valdez oil spill, a ship carrying BP oil ran aground and spilled at Huntington Beach, California. The BP CEO flew to the spill, and had obviously planned his risk communication carefully. When he was asked, 'Whose fault was this spill?' you could see he wanted to say 'Look, it was a contract ship, with a contract crew. They spilled our oil!' But instead he said: 'My lawyers say this was not our fault, but I feel as if it were our fault, and we will deal with it as if it were our fault.' Six months after the spill, they polled the residents, and BP had a higher approval rating than before the spill.

This 'responsive process' of communication has four facets (Snow 1997):

1. Openness and disclosure;
2. Acknowledgement of wrongdoing;
3. Courtesy (even if public response is angry and impolite);
4. Compassion (recognising and addressing people's fears and apprehensions).

According to Sandman, 'Risk communication comprises two facets: "scaring people" and "calming people down", or alerting and reassuring people. There are moderate hazards that people are apathetic about or minor hazards that people are outraged about.' The goal of risk communication is to create a level of outrage appropriate to the level of hazard (Snow 1997).

Since the public responds more to outrage than to hazard, 'Risk managers must work to make serious hazards more outrageous, and modest hazards less outrageous.' Stoking the outrage has been a successful strategy in increasing public concerns about the serious hazards of drunk driving for example (Sandman 1987). Conversely, any strategy that attempts to decrease public concern about minimal hazard must decrease the outrage. Sandman states, 'When people are treated with fairness and honesty and respect for their right to make their own decisions, they are a lot less likely to overestimate small hazards. At that point risk communication can help explain the hazard. But when people are not treated with fairness and honesty and respect for their right to make their own decisions, there is little risk communication can do to keep them from raising hell – regardless of the extent of the hazard.'

Any agency responsible for conveying risk information to the public should consider using the services of professional risk communicators. Adams and Sachs (1991) suggest that '"active listening" may be a more important component of effective communication than "telling". People are more open to hearing facts, or data, if they feel they have been listened to' (Adams and Sachs 1991). They also describe community outreach programmes instituted by the US Department of Agriculture Food Safety and Inspection Services, which were designed to interface with the public in such a way as to obtain 'advice, comments, and recommendations from consumers, industry, public-interest groups, our labour union, and our inspectors'. To prevent panic in the face of a potential food scare, sources of information must be trusted. The first source of information is always the government: Government agencies, public health departments, government spokespersons and so on. Unfortunately, as discussed above, they are the least trusted. Neither is industry trusted to tell the truth about hazards they are responsible for. On the other hand, university scientists and medical groups are the most trusted sources. If you seek to reassure, then use them to convey the message. A message is more believable when it comes from a totally unexpected source. If government or industry says something is harmless, it doesn't ring true for the public, since no one expects them to tell the truth. On the other hand, since industry is expected to down-play or cover up hazards, if they say it's dangerous, they will be trusted on that issue. And, since it's the job of an environment group or a consumer group to sound the alarm about hazards, if they say there's no danger, they'll be trusted as well. The source of the message is often more important than the content of the message.

Conclusions

Governments and regulatory authorities have become over-protective. It is time to redress the balance by placing more responsibility on individuals for the consequences of their own actions. Mistakes made by regulators are not learning experiences, as they would be for an individual, but causes of recrimination and resentment. Perhaps the following label on a USA food product may be more effective than all the regulatory requirements together!

Safe Handling Instructions

This product was prepared from inspected and passed meat or poultry. Some food products contain bacteria that could cause illness if the product is mishandled or cooked improperly. For your protection follow these safe handling instructions.

Keep refrigerated or frozen. Thaw in refrigerator or microwave.

Keep raw meat and poultry separated from other foods. Wash working surfaces (including cutting boards) utensils and hands after touching raw meat.

Cook thoroughly

Keep hot foods hot.

Refrigerate leftovers immediately or discard.

Figure 5.4 USA food warning label

References

Adams, C. E., Sachs, S. (1991). Government's Role in Communicating Food Safety Information to the Public, *Food Technology*, May, pp. 254–255.
Axelrad, J. (1998). An Auto-immune Response Causes Transmissible Spongiform Encephalopathies, *Medical Hypotheses*, **50**, 259–264.
Begley, S. (1991). The Contrarian Press, *Food Technology*, May 245–246.
Booker, C. and North, R. (1994). *The Mad Officials*, London, Constable.
Bowbrick, J. (1977). The Case Against Compulsory Minimum Standards, *Journal of Agricultural Economics*, **28**, 113.
British Medical Association (BMA) Guide, (1987). *Living With Risk*, J. Wiley and Sons.
Bruce, M. E., Will, R. G., Ironside, J. W., et al. (1997). Transmissions to Mice Indicate that 'New Variant' CJD is Caused by the BSE Agent, *Nature*, **389**, 498–501.
Burke, T. and Shackleton, J. R. (1996). *Trouble in Store*, Institute of Economic Affairs, Hobart Paper No. 130, London.
Cannon, P. (1987). *The Politics of Food*, London: Century, p. 314.
Communicable Disease Report (1996), Food Poisoning: notifications, laboratory reports, and outbreaks – where do the statistics come from and what do they mean?, 6, Review No. **7**, 21 June.
Craven, B. M. and Stewart, G. T, (1997). 'Public Policy and Public Health: coping with potential medical disaster' in: *What Risk?* Ed. R. Bate, Butterworth-Heinemann, Oxford.
Craven, B. M., Stewart, G. T. and Taghavi, M. (1994). 'Amateurs Confronting Politicians: A case Study of AIDS in England', *Journal of Public Policy*, **13**, 4:305–325.
Daily Telegraph (1998). 'Stand Up to the Use-by Bullies' 7 March.
Downs, A. (1957). *An Economic Theory of Democracy*, Harper and Row.
Downs, A. (1967). *Inside Bureaucracy*, Little, Brown and Co. Boston.
Fife-Schaw, C., Rowe, G. (1996). Public Perceptions of Everyday Food Hazards: A psychometric study, *Risk Analysis*, **16**, 487–500.
Financial Times (1994). 'Through Gas and High Water', 30 May.
Frewer, L. J., Howard, C., Hedderley, D., et al. (1996). What Determines Trust in Information about Food-related Risks? Underlying psychological constructs, *Risk Analysis*, **16**, 473–486.
Frewer, L. J., Howard, C., Hedderley, D. et al. (1997). The Elaboration Likelihood Model and Communication about Food Risks, *Risk Analysis*, **17**, 759–770.
Galbraith, J. K. (1973). *Economics and the Public Purpose*, Houghton Mifflin.

Groth, E. (1991). Communicating With Consumers about Food Safety and Risk Issues, *Food Technology*, May, 248–253.
Halsbury's Statutory Instruments, (1996). Vol. 8 1996 Re-issue, London, Butterworths.
Harveys, I. (1991). 'Infectious Disease Notification. A neglected legal requirement', *Health Trends*, **23**, 73–74.
Hill, A. F., Desbruslais, M., Joiner, S., et al. (1997). The Same Prion Strain Causes nvCJD and BSE, *Nature*, **389**, 448–450.
Hobbs, B. C. and Roberts, D. (1990). *Poisoning and Food Hygiene*, Fifth Edition, London.
Kay, J. (1988). 'The Economics of Regulation' in Seldon, A. (ed.), *Financial Regulation or Over-Regulation?*, Institute of Economic Affairs, pp 17–31.
Kone, D., Mullet, E. (1994). Societal Risk Perception and Media Coverage, *Risk Analysis*, **14**, 21–24.
Lave, L. (1987). Health and Safety Risk Analyses: Information for better decisions, *Science*, **236**, 291–295.
Lee, K. (1989). Food neophobia: major causes and treatments, *Food Technology*, December, 62–73.
Maitland, A., (1998). '"One in a Billion Risk" from Beef on the Bone', *Financial Times*, 10 February 1998.
Morabia, A., Porta, M. (1998). Ethics of Ignorance: lessons from the epidemiological assessment of the bovine spongiform encephalopathy ('mad cow disease') epidemic, *Perspectives in Biology and Medicine*, Winter 1998, 259–266.
NBPA (National Back Pain Association), Statistical sources: Department of Social Security, Department of Health, Health and Safety Executive, Clinical Standards Advisory Group, OPCS, RCN.
Niskanen, W. (1971). *Bureaucracy and Representative Government*, Aldine, Chicago.
Niskanen, W. (1973). *Bureaucracy: Servant or Master?*, Hobart Paperback 5, Institute of Economic Affairs, London.
North, R. (1992). *Death by Regulation: the butchery of the British meat industry*, Institute of Economic Affairs Health and Welfare Unit, Health Series, No. 12.
North, R. and Gorman, T. (1990). *Chickengate: An Independent Investigation of the Salmonellæ in Eggs Scare*, Institute of Economic Affairs Health and Welfare Unit paper No. 10. London.
Posner, R. A. (1974). Theories of Economic Regulation, *Bell Journal of Economics and Management Science*, **5**, 335–358.

Roberts, J. A. and Sockett, P. N. (1994). 'The Socio-economic Impact of Human *Salmonellæ enteritidis* Infection', *International Journal of Food Microbiology*, **21**, 117–129.
Sandman, P. M. (1994). Mass Media and Environmental Risk: Seven principles, *RISK: Health, Safety, and Environment*, Summer, 151–260.
Sandman, P. M. (1987). Risk Communication: facing public outrage, *EPA Journal*, November, 21–22.
Schefferle, H. (1990). 'Legislation' in Hobbs, B. C. and Roberts, D. *Poisoning and Food Hygiene*, Fifth Edition, London.
Scott, C. (1990). 'Continuity and Change in British Food Law', *Modern Law Review*, **53**, 785.
Seddon, Q. (1990). *Spoiled for Choice: Food Scares Unscrambled*, Evergreen Publishing, Essex.
Skegg, D. C. (1997). Epidemic or False Alarm?, *Nature*, **385**, 200.
Slovic, P., Fischhoff, B., Lichtenstein, S. (1981). Perceived Risk: psychological factors and social implications, *Procedures of the Royal Soc. London*, **376**, 17–34.
Snow, E. (1997). *Risk Communication*: notes from a class by Dr. Peter Sandman. Found at web site http://www.owt.com/users/snowtao/risk.html.
Social Trends (1997). Central Statistical Office, HMSO, London.
Sparks, P. and Shepherd, R. (1994). Public Perceptions of the Potential Hazards Associated with Food Production and Food Consumption: an empirical study, *Risk Analysis*, **14**, 799–806.
Stigler, G. J. (1971). The Theory of Economic Regulation, *Bell Journal of Economics and Management Science*, **2**: 3–21.
Thompson, K. (1996). *The Law of Food and Drink*, Shaw and Sons, Crayford, Kent.
Tower, G. (1993). A Public Accountability Model of Accounting Regulation, *British Accounting Review*, **25**, 1: 61–85.
Veljanovski, C. (1991). *Regulators and the Market*, Institute of Economic Affairs.
Voss, S. (1992). How Much do Doctors Know About the Notification of Infectious Diseases? *British Medical Journal*, **304**, 726–727.
Wildavsky, A. (1973). *Does Planning Work?* The Public Interest, No. 33.
Wildavsky, A. (1975). *Budgeting: A Comparative Theory of Budgetary Processes*, Little, Brown and Co. Boston.
Williamson, O. E. (1975). *Markets and Hierarchies: Analysis and Anti-Trust Implications*, Free Press, New York.

6 Dietary nitrates pose no threat to human health

Jean-Louis L'hirondel

Summary

Two major charges were levelled at nitrates some thirty years ago: infant methaemoglobinaemia or 'blue-baby syndrome', and a greater risk of cancer in adults. These were either presumptions or hypotheses.

The many scientific studies carried out over the last few decades allow us to conclude that neither of those grievances were founded. Dietary nitrates pose no threat to human health.

Nitrates present in blood plasma have two sources; one source of nitrates is exogenous – from food; 80 per cent of these alimentary nitrates come from vegetables and 10 to 15 per cent from drinking water; there is also an endogenous source, providing a similar quantity, of cell origin, involving the amino acid, L-arginine and nitrogen monoxide (NO). In addition to passive urinary excretion, there are two active secretions of nitrate (NO_3) from plasma: colonic secretion and salivary secretion. Salivary secretion reintroduces NO_3 ions a second time in the mouth.

The directives issued in 1962 by the UN's WHO and Food and Agricultural Organisation (FAO), and in 1980 by the EEC are now redundant. The directive on drinking water is in addition very costly. They need repealing; eventually, this will become inevitable.

Introduction

From the twelfth to the nineteenth century, nitrates (NO_3) were used as medicines, sometimes in very large doses, for a wide, and sometimes surprising, range of symptoms.

At the beginning of the twentieth century, the development of aspirin followed by the introduction of corticoids meant the end of the therapeutic use of nitrates. By the 1950s, the growing incidence of infant methaemoglobinaemia in some rural areas of the United

States, together with the discovery of the carcinogenicity of many nitrosamines in animals, contributed to bringing dietary nitrates under suspicion and to casting doubt on their innoccuousness.

In 1962 these doubts led the Committee of Experts on Food Additives of the WHO and the FAO to set an acceptable daily intake (ADI) level for man at 3.65 mg/kg of NO_3, and in 1980 the European Community issued a directive setting a limit for NO_3 in drinking water of 50 mg/l, above which water is no longer deemed fit for human consumption.

Some thirty-six years after the WHO and the FAO's decision and eighteen years after the European Community's directive, the scientific perspective has changed totally. Many studies and experiments have taken place; any suspicions which were legitimate a few decades ago can no longer be justified.

The metabolism of nitrates

Figure 6.1 gives an overview of the metabolism of nitrates. Nitrates are always present in our bloodstream, at levels normally ranging between 1 and 3 mg/l before meals. In normal conditions two sources of nitrates coexist – the exogenous source from food and water, and the endogenous source from cell activity; they each provide 70 to 75 mg/day.

Of the food source, 80 per cent of nitrates ingested by man normally come from vegetables, and 10 to 15 per cent from drinking water. On swallowing, these nitrates pass down into the stomach as NO_3, before being quickly and virtually totally absorbed in the upper section of the small intestine. Less than 2 per cent of all nitrates ingested therefore reach the large intestine (Bartholomew 1984).

The endogenous source of nitrates has been known for some thirteen years, since the research carried out by Stuehr and Marletta in 1985. The metabolic process of the amino acid L-arginine releases a nitrogen atom at cell level which forms a molecule of nitrogen monoxide, NO. Outside the cell, the NO molecules combine with oxygen to form various molecules including nitrates, nitrites and nitrosamines. Many physiological activities, like running and cycling, and pathological conditions, like infections, lead to cell stimulation and thereby contribute to increasing this endogenous synthesis of nitrates.

Thereafter, the fate of plasmatic nitrates, from two different sources – exogenous and endogenous – is rather unusual:

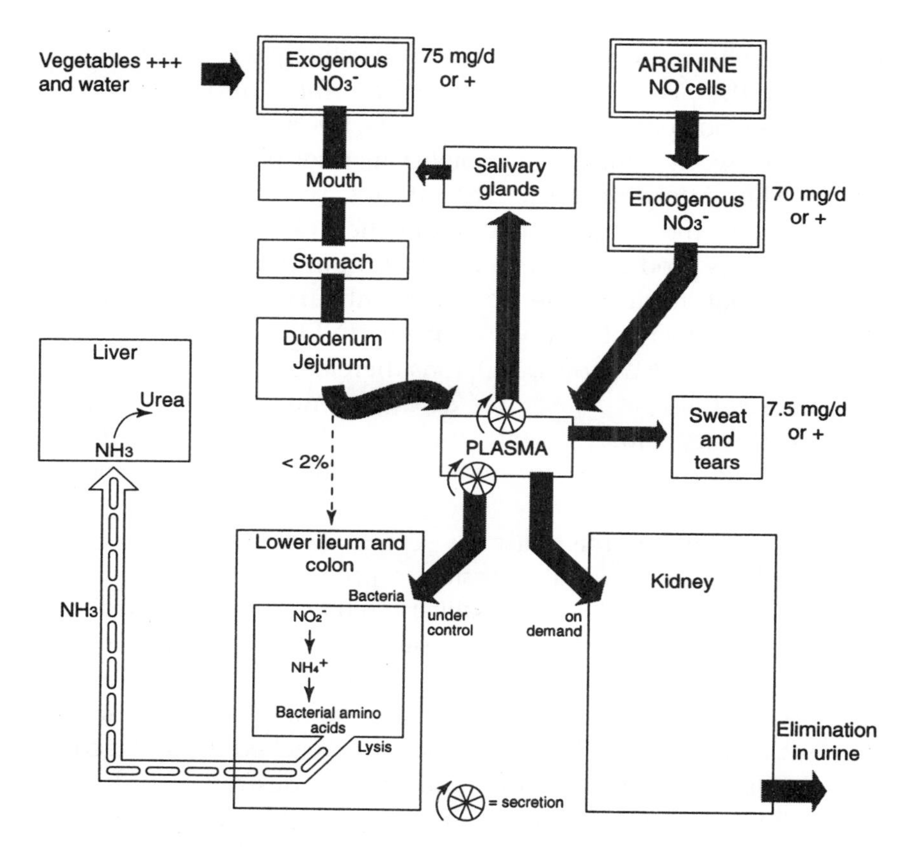

Figure 6.1 The metabolism of nitrates in man

- A small proportion of plasmatic nitrates, approximately 10 per cent of the quantity of nitrates ingested, is eliminated through sweat and tears.
- The largest amount of nitrates leaving the plasma is passively excreted in urine, this excretion being based only on NO_3 plasmatic concentrations.
- Two active phenomena also occur, which have a significant impact from a physiological viewpoint: the colonic and salivary secretions of nitrates.
- One of the functions of the columnar cells of the colonic epithelium is to draw NO_3 ions from the plasmatic sector towards the colon's light through an active capture phenomenon. The purpose of this colonic secretion of NO_3 ions is most probably to ensure the nutrition of the colonic bacterial flora.

- The cells of the salivary acinus also actively draw NO_3 ions from the plasmatic sector and release them into the saliva, its secretion product. At times therefore, levels of salivary nitrates are between 6 and 30 times higher than for plasmatic nitrates.

  Thereafter, these salivary nitrates, which remain for a while in the mouth, come under the influence of bacterial enzymes produced from a relatively abundant physiological bacterial flora. Some of these salivary nitrates (NO_3) thereby turn into salivary nitrites (NO_2).

  As Figure 6.2 shows, NO_3 ions therefore pass through the mouth twice, the first time as dietary nitrates, the second as salivary nitrates. Only the latter process induces the formation of a certain amount of salivary nitrites, which reach the stomach when the saliva is swallowed.

  The role of this salivary secretion of nitrates merits clarification. It is quite possibly a preliminary stage in the digestion of proteins, as salivary nitrates have the ability to make food proteins more sensitive to the subsequent action of proteolytic enzymes (pepsine and trypsine). Benjamin also showed in 1994 that in an acid medium, ingested salivary nitrates, which have therefore reached the stomach, release nitrogen monoxide and thereby destroy organisms such as **Candida albicans** and **Escherichia coli**, thereby promoting host defence against ingested pathogens.

Grievances against nitrates and their refutation

In the 1950s and 1960s, alimentary nitrates aroused disquiet on two counts. They were thought to be responsible for methaemoglobinaemia in infants, and people wondered whether they might not also induce the onset of some cancers, in particular stomach cancer. These two presumptions gave rise to two major grievances. The many studies conducted in the last thirty years now allow us to state that both are unfounded.

Dietary methaemoglobinaemia in infants – Blue Baby syndrome

Methaemoglobin is an oxidised derivative of haemoglobin, which loses its ability to carry oxygen molecules. In the physiological

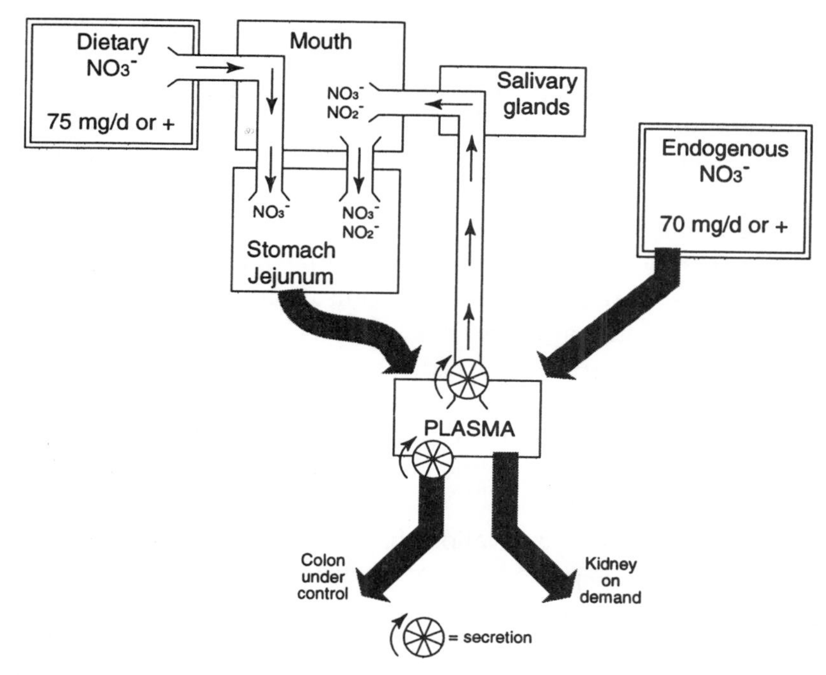

Figure 6.2 The duality of nitrates in the mouth. Dietary nitrates and nitrates secreted by the salivary glands are two distinct entities. Only NO_3^- ions secreted by the salivary glands, as NO_2^- precursors, can induce nitrosamines. By contrast, NO_3^- ions from food reach the stomach intact, intact, i.e. without turning into NO_2^- and therefore without any risk of turning into nitrosamines

state, 1 to 2 per cent of haemoglobin in the red cells are in the form of methaemoglobin. Clinical disorders, in this case cyanosis, appear if methaemoglobinaemia levels exceed the 10 to 20 per cent threshold. Any increase reaching 70 to 80 per cent can be fatal.

The transformation of haemoglobin into methaemoglobin in red cells is due to the action of nitrites rather than nitrates, which are very powerful oxidants. Babies under the age of six months are most at risk as they have not yet fully developed a protective enzyme system (reductase methaemoglobin or NADH-cytochrome b5 reductase). Beyond the age of six months, the risk of pathological methaemoglobinaemia no longer exists.

Prior to 1984, it was generally thought that nitrates in feeding bottles turned into nitrites in the baby's colon, after contact of these nitrates with the colon's large bacterial flora, as only bacterial enzymes are capable of reducing NO_3 nitrates into NO_2 nitrites. However, since Bartholomew's studies in 1984 showed that 98 per cent of alimentary nitrates are absorbed in the upper section of the small intestine, this explanation is no longer valid.

Some scientists then wondered whether the nitrates–nitrites transformation could not take place in the baby's stomach, as a result of a colonisation of the stomach by micro-organisms of enteric origin, under a hypochlorhydric effect of the gastric juices. However, studies on this subject are not conclusive, as Walker showed in 1990; the secretion of gastric acid in infants is actually sufficient to prevent any significant bacterial colonisation.

In reality, as some authors (Knotek 1964, Simon 1966, Dupeyron 1970, J. L'hirondel 1971) had indicated twenty-five to thirty years ago, the nitrates–nitrites transformation, responsible for infant methaemoglobinaemia, occurs in the feeding bottles, if basic hygiene rules are not observed when the bottles are being prepared, thereby causing microbe pullulation.

A number of clinical observations have confirmed this; these include the sudden and unexpected nature of cyanosis, when large amounts of nitrated foods (carrot soup, spinach) have been ingested during the days or weeks preceding the incident without giving rise to the slightest clinical anomaly, the rapid onset of cyanosis 15 to 20 minutes after feeding, and the lack of correlation between the level of methaemoglobinaemia and the amount of nitrates ingested with food.

To limit nitrate levels at 50 mg/l of NO_3 in drinking water does not constitute an appropriate preventive response with regard to infant methaemoglobinaemia.

Nitrates will continue to be a potential element of infant food via the intake of water used to prepare formula milk, and vegetables. The only effective preventive solution consists in acting on the bacterial element: all risks of bacteria pullulation must be avoided in feeding bottles, whether they contain nitrates or not, by following a few basic hygiene rules when the bottles are being prepared. In the case of carrot soup, it should be boiled for a few minutes and fed to the baby shortly thereafter, and in no circumstances should the soup be left to stand at ambient temperature for more than six to eight hours.

What causes methaemoglobinaemia in infants therefore is not alimentary nitrates, but rather the nitrites formed in the feeding bottles after the reduction of nitrates into nitrites as a result of an unfortunate microbe pullulation in the bottles. It is the latter phenomenon which has to be prevented at all cost. Infant methaemoglobinaemia caused by food has been virtually eradicated in developed countries, where people are familiar with basic hygiene rules for preparing bottles of formula milk.

Cancer

As shown above, the salivary glands draw nitrate ions from the plasmatic sector; salivary nitrates partly turn into nitrites in the mouth, and, on swallowing, these salivary nitrites reach the stomach.

Salivary nitrites then react with various amines in the stomach to form nitrosamines. Ninety per cent of nitrosamines tested in experiments are known to be carcinogens in animals. It was deduced from this that nitrates have a potential carcinogenic power, and this presumption has now been hanging over them for almost forty years.

On analysis, however, this suspicion proves unfounded:

1. As Figures 6.1 and 6.2 show, nitrites in the stomach do not come directly from alimentary nitrates; they come from plasmatic nitrates on which the salivary glands have had a very specific action.
2. The amount of nitrosamines thereby formed in the stomach through the metabolism of nitrates is very tiny.

Figure 6.3 compares levels between nitrosamines produced from endogenous synthesis, i.e. via the body's cells through the metabolic process involving L-arginine, those brought in by direct exogenous intake via food, those brought in exogenously by extra-dietary means (tobacco, tyre factories, tanneries), and the no observed-adverse effect level, in animals.

Many foods (beer and seasoned cooked meat in particular) contain nitrosamines. Levels of intake by direct dietary means are several tens or hundreds of times higher than for nitrosamines formed in the stomach through the metabolism of nitrates. If the precautionary principle were to be used, it would have to be applied in priority to nitrosamines of direct dietary origin, which would involve introducing restrictive measures for a number of foods. In fact, such restrictive measures are not necessary as these endogenous and exogenous nitrosamines remain confined to very tiny amounts compared to the theoretical toxicity threshold.

The level of direct dietary nitrosamines intake is several hundreds of times lower than the potential toxic level, and the amount of nitrosamines formed in the stomach during the metabolism of

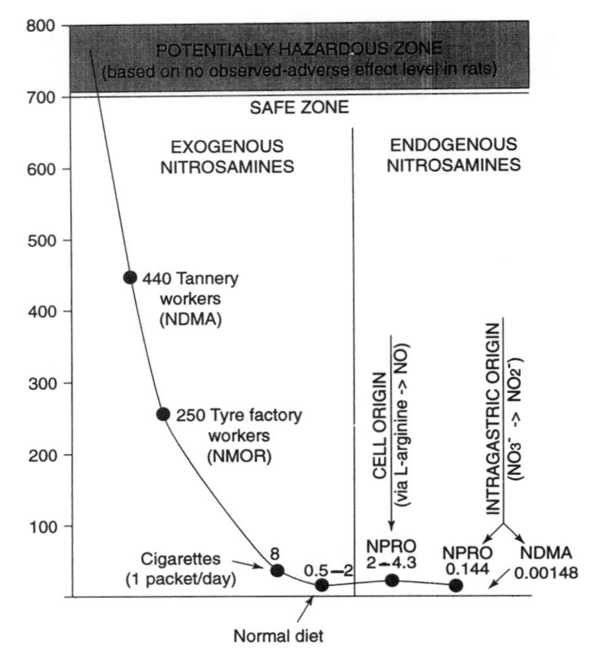

Figure 6.3 Comparison of endogenous syntheses and exogenous intakes of nitrosamines. NDMA: nitrosodimethylamine, NMOR: nitrosomorpholine, NPRO: nitrosoproline

nitrates is several tens of thousands of times smaller than the potential toxic level (J. L'hirondel and J. L. L'hirondel 1996).

3. All nitrates–cancer experimental studies have proved negative in animals. Not one study conducted on rats or mice has succeeded in showing that even a considerable and prolonged intake of nitrates results in an increase in the incidence of cancers.
4. Since 1945, some twenty epidemiological studies have attempted to clarify the possible correlation in man between nitrate intake and the incidence of stomach cancers. Only two out of twenty show a positive correlation. Seven out of twenty even point to a statistically significant negative correlation. Such a negative correlation should not surprise us at all; the favourable impact of vegetables on the incidence of cancer pathology in general is universally acknowledged (WHO, 1990), and, as we know, 80 per cent of ingested nitrates come from vegetables.

> Therefore, as stated by the European Commission's Scientific Committee for Food in its 'Opinion on Nitrate and Nitrite' (22nd September 1995): 'Epidemiological studies thus far have failed to provide evidence of a causal association between nitrate exposure and human cancer risk'.

In conclusion therefore, the amounts of nitrosamines formed in the stomach during the metabolism of nitrates are actually very tiny; in no way are they capable of increasing the incidence of cancer pathology in man.

Other grievances

Other, less serious charges have been levelled at dietary nitrates: an increase in the risk of foetal death, an increase in the risk of congenital malformation, a tendency towards enlargement of the thyroid gland, and an early onset of arterial hypertension.

There have been few studies on these issues, and some of them contain a number of methodological flaws. As a result, these ancillary grievances cannot be sustained legitimately as they lack a sound, documented scientific basis.

Conclusions

Whether they are considered major or secondary therefore, no grievances against dietary nitrates in food can stand up to analysis. Scientific knowledge leads to the following conclusion: in the short, medium and long term, nitrates from food and from drinking water have no negative impact on human health.

Consequently, the directive drawn up in 1962 by the Committee of Experts on Food Additives of the WHO and the FAO on an acceptable daily intake (ADI) level for man is now redundant; so is the directive from the European Community dated 15th July 1980 on the quality of water intended for human consumption (80/778/EEC), which set a permitted limit of 50 mg/l for NO_3, above which water is no longer deemed fit for human consumption.

The implementation of this latter directive on drinking water is particularly costly for the citizens of the European Community. Its repeal is necessary and inevitable.

References

Bartholomew, B., Hill, M. J. (1984). The pharmacology of dietary nitrate and the origin of urinary nitrate, *Fd. Chem. Toxic.*, **22**, 789–85.

Benjamin, N., O'Driscoll, F., Dougall, H. et al. (1994). Stomach NO synthesis, *Nature*, **368**, 502.

Dupeyron, J. P., Monier, J. P., Fabiamni, P. (1970) Nitrites alirnentaires et methemoglobinemie du nourrisson, *Ann. Biol. Clin.*, **28**, 331–6.

European Commission (1980). Directive on the quality of drinking water for human consumption, Council Directive 80/778/EEC OJNI L229, 30-8-1980, 11–26.

European Commission (1995). Scientific Committee for Food Annex 4 to document III/15611/95. Opinion on nitrate and nitrite (expressed on 22 September 1995), 20.

Knotek, Z., Schmidt, P. (1964). Pathogenesis, incidence and possibilities of preventing alimentary nitrate methemoglobinemia in infants, *Pediatrics*, **34**, 78.

L'Hirondel, J., Guihard J., Morel, C. et al. (1971). Une cause nouvelle de méthémoglobinémie du nourrisson: la soupe des carrottes, *Ann. Pediat.*, **18**, 625–32.

L'hirondel, J., L'hirondel J. L. (1996). *Les nitrates et l'homme, le mythe de leur toxicité.* Editions de l'Institut le 1'Environnement, 142 p.

Simon, C. (1966). L'intoxication par les nitrites après ingestion d'épinards (une forme de méthémoglobinémie) *Arch. Fr. Ped.*, **23**, 231–8.

Stuehr, D. J., Marletta, M. A. (1985). Mammalian nitrate biosynthesis: Mouse macrophages produce nitrite and nitrate in response to *Escherichia coli* lipolysaccharide, *Proc. Nati. Acad. Sci.*, **82**, 7738–42

Walker, R. (1990). Nitrates, nitrites and N-nitrosocompounds: a review of the occurrence in food and diet and the toxicological implications, *Food Add. Contam.* **7**, 717–68.

WHO (1962). Evaluation of the toxicity of a number of antimicrobials and antioxidants. Sixth report of the joint FAO/WHO Expert Committee on Food Additives. World Health Organization, Technical Report Series 228, 76–78, Geneva.

WHO (1990). Régime alimentaire, nutrition et prévention des maladies chroniques. Rapport d'un groupe d'étude de 1'OMS. Série de rapports techniques 797.

7 Rachel's folly: The end of chlorine*

Michelle Malkin and Michael Fumento

Summary

Millions of people are dying all over the world because of dirty water. Radical environmentalists have tried to claim that the main culprits are man-made pollution and chlorinated chemicals. The more likely culprits are endemic poverty, bad plumbing and lack of access to basic water chlorination techniques. According to the World Health Organisation, every year, nearly 1.5 billion people – mostly children under five – suffer from preventable water-borne diseases such as cholera, typhoid fever, amoebic dysentery, bacterial gastro-enteritis, giardiasis, schistosomiasis, and various viral diseases, such as hepatitis A (WHO 1998). Yet now there is a mounting campaign, led by environmental activists in wealthy industrialised nations, to eliminate every last man-made chlorine molecule from the face of the earth.

Environmentalist attack on chlorine

Greenpeace, the international environmental advocacy group, launched the first salvo in 1991 with its call to phase out completely 'the use, export, and import of all organochlorines, elemental chlorine, and chlorinated oxidising agents (e.g. chlorine dioxide and sodium hypochlorite)' (Thornton 1991). Yet chlorination – considered one of the greatest advances ever in public health and hygiene – is almost universally accepted as the method of choice for purifying water supplies (IARC 1991). In the United States alone, 98 per cent of public water systems are purified by chlorine or chlorine-based products. Alternative chemical disinfectants such as ozone and other short-lived free radicals have been used in water treatment, but none has demonstrated the safety and efficacy of chlorination (Mintz 1995).

* An earlier version of this chapter was published by the Comparative Enterprise Institute, Washington DC in 1996.

Chlorine is a ubiquitous element, one of the basic building blocks of all matter in the universe. In fact, scientists are only now beginning to discover and identify the great number of natural organohalogens present in our world. By one estimate, Mother Nature manufactures at least 1500 chlorine-containing chemicals (Gribble 1994). Volcanic activity, forest and grass fires, fungi, algae, ferns and the decomposition of seaweed all release chlorinated organics into the environment (Willes 1993). Our own bodies produce hypochlorite to fight infection and hydrochloric acid for proper digestion (Gribble 1995). And there is, of course, sodium chloride – common table salt – present naturally in mines, lakes and sea water, found in our blood, sweat and tears, and essential to the diets of humans and animals (Winterton 1997).

Clearly, a goal of total chlorine removal from the environment would be unattainable, and the potential human toll resulting from its eradication is manifest and staggering. Every major scientific investigation of chlorinated water has concluded that the real and proven health risks from microbial contamination of drinking water far exceed the uncertain hypothetical risks of cancer from chlorination and its by-products. Why, then are governmental bodies around the world embracing Greenpeace's caprice – absolute zero tolerance for man-made chlorine – when the hazards to humanity are so explicitly large?

Perhaps the answer can be traced back to the publication of Rachel Carson's *Silent Spring* in 1962. The book is a lyrical tract, the bible of the environmental movement. Carson was the first to bear witness against chlorinated hydrocarbons and other 'elixirs of death' created by 'the ingenious laboratory manipulation of molecules' (Carson 1984). She condemned these arrogant manipulations, prophesied a man-made cancer epidemic, and popularised the zero-based approach to regulating synthetic chemicals. A daunting theme runs throughout *Silent Spring* – that man's ingenuity would be his own worst enemy. And therein lies the essence of Rachel's folly. Carson and her intellectual heirs in the environmental movement embrace a mistaken vision of technology. It is an impaired vision that considers only the risks of industrial chemical compounds, and not the risks created by their absence.

As the late Aaron Wildavsky observed, there are few unalloyed good things in the world. Rarely does one find a substance that has the benefits but not the costs (Wildavsky 1988). 'Sunsetting' all uses of chlorine may reduce the hypothetical risks associated with such compounds as dioxin, DDT and PCBs. At the same time, however, a blanket ban on chlorine would increase the enormous risks of water-borne microbial infection here and in underdeveloped countries that can now barely afford chlorine disinfectants (let alone costly substitutes such as ozone or ultraviolet light treatment). (See also Ghersi and Ñaupari, Chapter Two.)

Even more daunting, a chlorine phase-out would halt the production of most plastics, pesticides and chlorine-containing drugs like chloroquine, a key anti-malarial drug; halogenated tetracycline-based antibiotics like chlortetracycline; and the family of halogenated anti-psychotics such as chlorpromazine (McCarty 1994). According to one industry-backed report, almost 85 per cent of the pharmaceuticals manufactured worldwide require chlorine at some stage of production; 96 per cent of crop-protection chemicals are chlorine-dependent (Charles River Associates 1993, Hilleman, 1993). From safe drinking water, clean swimming-pools and pest-free crops, to flame retardants and food packaging, quality white paper and bright socks, food wrap, plastic bottles, garden hoses, window frames and sturdy plumbing pipes, the end of chlorine would have far-reaching, unintended consequences.

Chlorine in the time of cholera

There is no plainer example of the health benefits of chlorine, and the health risks of its absence, than the cholera epidemic in Latin America. In February 1991, the first cholera outbreak to hit Peru since the turn of the century was reported (*The Lancet* 1992). According to the journal *Nature*, US and international health officials blamed the occurrence on Peruvian government officials who made a 'gross miscalculation' in not chlorinating the water supply (Anderson 1991).

It is believed that local water officials in Lima were influenced in their decision to reduce chlorination in many of the wells because US Environmental Protection Agency (EPA) studies conducted in the mid-1980s showed an increased hypothetical cancer risk from trihalomethanes (THM), a chlorination by-product. One of those studies (based on high-dose experiments on animals exposed over their lifetimes) estimated a risk of up to 700 additional cancer cases per year in the US from THMs; by contrast, however, the Latin American cholera epidemic claimed nearly 4000 lives in 1991 alone (*The Lancet* 1992).

EPA administrators denied that risk communication failures on their part could be faulted for touching off the epidemic. Many American researchers, however, questioned whether EPA should have given more emphasis to the disaster potential of not disinfecting municipal water supplies (Anderson 1991). Whatever the actual impact EPA calculations had in Lima, a follow-up study in Peru's second largest city, Trujillo, pointed to the two bottom-line causes of the outbreak and its rapid spread. Quite simply, they were a lack of chlorinators and a shortage of funds to buy them (Swerdlow et al. 1992).

Preliminary data examined by Mintz et al. (1995) suggest that intervention costs for point-of-use disinfection in developing countries is low: 'The annual cost per family for both a special water storage vessel and (chlorinated) disinfectant, for the shortest estimated useful life of the vessel and the highest cost of hypochlorite, would be between $1.17 and $1.62, an amount affordable almost anywhere in the world.' In the March 1995 issue of the *Journal of the American Medical Association*, the researchers endorsed the expanded use of sodium and calcium hypochlorite – deemed 'relatively safe, easy to distribute and use, inexpensive, and effective against most bacterial and viral pathogens' – to prevent persistent waterborne disease (Mintz et al. 1995). In addition to cholera, these infectious diseases include typhoid fever, amoebic dysentery, bacterial gastro-enteritis, shigellosis, salmonellosis, Campylobacter enteritis, Yersina enteritis, Pseudomonas infections, schistosomiasis, giardiasis and various viral afflictions, such as hepatitis A.

In a campaign to increase access to potable water in poor countries, the World Health Organisation (WHO) declared the 1980s the 'Drinking Water Supply and Sanitation Decade'. Access to proper chlorination, however, remains a major barrier and efforts to improve both municipal water treatment and home storage techniques continue. At last count, the WHO estimated that 25 million people – 70 000 per day, mostly children under five – die around the world each year from dirty drinking water. While non-chlorine disinfectants like iodine, ozone, and short-lived free radicals have been used to treat water on a limited basis, none has demonstrated the safety and cost-effectiveness of chlorination (Mintz et al. 1995).

The 'chlorine kills' campaign

As the Latin American cholera epidemic escalated, environmental activists a world away were building their arsenal against chlorine. Greenpeace, the international environmental advocacy group, launched the first salvo in early 1991 with its call to phase out completely 'the use, export, and import of all organochlorines, elemental chlorine, and chlorinated oxidising agents (e.g. chlorine dioxide and sodium hypochlorite)' (Amato 1993). As Greenpeace's Joe Thornton concluded, 'There are no uses of chlorine which we regard as safe' (Amato 1993).

Dismissing the proven benefits of clean chlorinated water (IARC 1991), the environmentalists focused attention instead on a short string of chlorine compounds such as polyvinyl chlorides (PVCs) used widely in modern consumer products and dioxin produced during bleaching by the pulp and paper industry. Greenpeace targeted the publishers of *Time* magazine and other major producers of chlorinated paper products from Canada to Ireland. They accused PVCs of being 'uniquely damaging during production, use and disposal', and they claimed that all chlorinated chemicals could cause cancer and reproductive damage. In short, as Greenpeace's Thornton explained, 'People should be considered innocent until proven guilty; chemicals should not' (Thornton 1991).

Trademark antics in the early days of the campaign included blockades of ships and trains carrying chlorine, and the unfurling of sensational, simplistic banners with slogans like 'Chlorine Kills!' and 'Take the Poison out of Paper', To underscore claims of environmental harm allegedly caused by chlorinated organic chemicals, activists sent a trawler called *Moby Dick* – aptly named after Melville's novel of a vengeful captain on an obsessive hunt – to tour the North American Great Lakes region in late summer 1991. *Moby Dick* took the anti-chlorine message to 40 cities; calls of a 'zero discharge' policy escalated. Soon the cover of *E, The*

Environmental Magazine, was asking 'Is Chlorine Killing the Great Lakes?'

Inside, the magazine quoted Barry Commoner, who likened the proliferation of chlorine compounds in modern life to the 'Invasion of the Body Snatchers': 'You can think of chlorinated biological compounds as aliens, and like aliens from outer space, the reason they cause problems is that they're readily assimilated into the normal chemistry of life. It's just like the movies' (Moberg 1993).

Initially, this extravagant rhetoric about organochlorines was limited to a few green Ahabs in the environment movement. But the concept of chlorine zero-discharge soon gained credibility with an endorsement from the venerable International Joint Commission (IJC).

The six-member IJC, a joint regulatory agency of non-scientists from the US and Canada, was created in 1909 to monitor commerce on, and assess water quality of, the Great Lakes. For most of its history, the panel refrained from alarmist policy pronouncements. In 1990, however, after emotional public hearings, demonstrations and intense lobbying efforts led by Greenpeace, the commission urged adoption of a 'binational toxic substances management strategy' based on the 'philosophy of zero-discharge' (IJC 1990). The panel's full-throttle endorsement of zero discharge was issued in 1992, when members called on both countries to 'develop time-tables to sunset the use of chlorine and chlorine-containing compounds as industrial feedstocks, and (examine) the means of reducing or eliminating other uses' (IJC 1992). Two years later, the IJC redoubled its call – the most far-reaching by any government body (IJC 1994).

What was the evidence that swayed the IJC to endorse a chlorine phase-out? Then-chairman Gordon Durnil, a lifelong Republican (with no prior scientific background) who was appointed to the commission by President George Bush in 1989, explained that he stayed up late at night reading studies to educate himself. He was appalled by various reports of breast cancer and reproductive harm in wildlife and humans that some scientists linked to industrial chemicals in the Great Lakes region. In his autobiography, Durnil confesses that, 'The truth is, in the beginning of my tenure, I wanted to disbelieve. But being a good conservative, with the ability to think for myself instead of being told how to think, I was willing to change my way of thinking. Evidence is evidence and facts are, indeed, facts' (Durnil 1995).

There is nothing wrong with putting capable non-scientists in charge of a fact-finding mission. But Durnil and his colleagues proved to be incapable of making the critical distinction between possibility of harm and probability of harm. They ignored studies that failed to bolster environmentalists' worst-case scenarios. They failed to consider the costs and risks associated with drastic regulatory action. And they succumbed to the environmental movement's bad habit of assuming harm, rather than assessing it. As a result, the IJC concluded that it had become 'necessary to shift the burden of responsibility for demonstrating whether substances should be allowed in commerce. The concept of reverse onus, or requiring proof that a substance is not toxic or persistent before use, should be the guiding philosophy of environmental management agencies in both countries...' (IJC 1994).

This endorsement of a mathematically and scientifically impossible standard of proof, that is, proving a negative, did not impress responsible toxicologists. Environmental activists, however, were ecstatic. As Greenpeace's Rick Hind, legislative director of the Toxics Campaign, told *Science* reporter Ivan Amato, 'The IJC lit our life' (Amato 1993).

Indeed, the IJC's 'sunset' proposal sparked governmental bodies and public health groups around the world to adopt zero tolerance policies that echoed Greenpeace's initial call to phase out chlorine. In September 1992, the Paris Convention of the North Sea – a European analogue of the IJC representing 13 nations – endorsed a ban on chlorine emissions in the north-east Atlantic Ocean. The Canadian provinces of Ontario and British Columbia enacted regulatory timetables to reduce chlorine use in the paper and pulp industry to zero by the year 2002.

In the US 1993 proved a watershed year for anti-chlorine activists. At the beginning of the year, the Clinton administration weighed in with a Clean Water Act initiative to develop a 'national strategy for substituting, reducing or prohibiting the use of chlorine and chlorinated compounds' (US EPA 1994). By the summer, Rep. Bill Richardson (D-NM) had offered up the 'Chlorine Zero Discharge Act of 1993'. Borrowing language used by the IJC (which was in turn borrowed from Greenpeace), the proposal called for a total phase-out of chlorine in the pulp and paper industry – in other words, 'absolutely no output or release, including non-point source output or release, into water' (US H.R. 1993).

Anti-chlorine activists were also buoyed by an endorsement from the American Public Health Association (APHA). In October 1993, the group of 30 000 public health professionals passed a divisive resolution calling for treatment of chlorine-containing organic compounds as a class. It recognised, as had the IJC, 'that the only feasible and prudent approach to eliminating the release and discharge of chlorinated organic chemicals and consequent exposure is to avoid the use of chlorine and its compounds in manufacturing processes' (APHA 1993). Echoing almost verbatim the Greenpeace philosophy, APHA spokesman Peter Orris declared that 'The APHA has found that the class of chlorine-containing chemicals should be considered guilty until proven innocent' (*BNA Chemical Regulation Daily* 1993).

Did the vote reflect the true assessment of APHA? The membership had considered the resolution once before – and rejected it; several years of in-fighting preceded adoption of the new position. The APHA leadership, like the IJC's, apparently succumbed to intense pressure from environmental lobbyists. In fact, Greenpeace publicised the APHA resolution after it was passed at the APHA's annual meeting in San Francisco (*BNA Chemical Regulation Daily* 1993).

Other environmental groups joined the clamour for a chlorine phase-out. The National Wildlife Federation, for example, called on the White House to support a sunset provision, and noted that 'the administration's commitment to promulgate a final strategy for substituting, reducing, or prohibiting the use of chlorinated compounds within two and a half years is critical' (BNA 1994). The Environmental Defense Fund, Friends of the Earth, National Audubon Society and others wrote to President Clinton to express their 'alarm and frustration' with the administration's delay on an executive order mandating government use of 'totally chlorine-free' paper (Letter 1993). And the Sierra Club called for 'immediate action to stop exposing men, women and children to these poisons. Regulation of [dioxin and other chlorinated] chemicals as a class is the only way that we can adequately address this issue' (Cox 1994).

In the spring of 1995, Rep. Richardson resurrected the Chlorine Zero Discharge Act with even more conviction: 'Federal intervention to ensure that the use of these unnecessary, dangerous chemicals is eliminated is needed now to protect the public from potentially life-threatening health and environmental impacts' (US Congressional Press Releases 1995). But just as the environmentalists seemed poised to score major legislative triumphs in the US, institutional momentum slowed. The proposed executive order mandating government use of 'totally chlorine-free' paper was withdrawn; the EPA's call for a chlorine study leading to product bans was put on hold. Richardson's proposal, H.R. 1400, is currently pending before the House Transportation Subcommittee on Water Resources and the Environment. Thirty-seven co-sponsors have signed on.

Meanwhile, the Latin American cholera epidemic entered its fourth year with nearly 1 million cases and 10 000 deaths directly attributable to dirty water and lack of chlorine disinfectants (*Weekly Epidemiol Rec.* 1993).

Stomach pains

In the chlorination process, chlorine reacts mainly with natural water constituents to produce a complex mixture of by-products, including a wide variety of halogenated compounds, the actual levels of which depend on the amount of chlorine added and the type of water source (IARC 1991). Besides trihalomethanes (which include chloroform, bromodichloro-methane, chlorodibromo-methane and bromoform), other chlorination by-products include halogenated acetic acids, chlorinated ketones and chlorinated furanones.

The idea that chlorination by-products in disinfected water cause cancer has gained widespread currency among the popular press and many environmentalists in wealthy developed countries. But the evidence of harm is weak. A review of studies published in the *Journal of the National Cancer Institute* found only equivocal evidence for carcinogenicity in female rats that received chlorinated or chloraminated drinking water seven days per week for two years in amounts ranging from 70 to 275 parts per million (ppm) and 50 to 200 ppm respectively; there was no evidence of carcinogenic activity in male rats or male and female mice administered the same amounts (Dunnick and Melnick 1993). By way of comparison, the EPA-regulated permissible residual level for chlorine in water is 1.5 ppm, and for chloramine, 4 ppm (US EPA 1988 1979).

While the Environmental Protection Agency has long contended that the carcinogenicity of chloroform (the most common THM) is substantial, the most up-to-date research has cast serious doubt on the agency's cancer estimates – and its primary method of inducing tumours in animals. Conventional EPA analysis suggests that an increased human risk of 1 in 100 000 can be expected from drinking water containing 4.3 parts per billion of chloroform over a lifetime. The basis for this estimate? Mouse experiments using a technique called 'gavage' in which large globs of chloroform dissolved in corn oil were pumped into the animals' forestomachs through a tube forced down the throat, five times a week over a 2-year lifespan.

Unsurprisingly, a group of private researchers discovered that cell damage due to the unusual administration of such high doses of chloroform caused excess liver cancer in the genetically tumour-prone mice. In the only animal studies of chloroform administered normally in drinking water, toxicologist Byron Butterworth et al. found no induced cellular proliferation in rodent livers even when the concentration of chloroform was over 1800 ppm – an amount several orders of magnitude above current regulatory limits (Larson et al. 1994). (The current total limit for THMs is 100 micrograms per litre; the occupational limit is 2 ppm for an 8-hour day.) As Butterworth noted, 'our studies thus far indicate no increased risk of cancer from the levels of chloroform found in drinking water' (Stone 1995). In scientific journals (but not the lay press), even the EPA conceded that their estimates were wrong. EPA toxicologist Rex Pegram admitted that 'Butterworth's work has gone a long way toward showing us that chloroform is not the worry it once was' (Stone 1995).

Indefinite exposure

Epidemiological evidence of harm from chlorinated by-products in drinking water is no more convincing. The International Agency for Research on Cancer (IARC) concluded, after exhaustive review, that epidemiological studies on the relationship between cancer and consumption of chlorinated drinking water were inadequate to draw definitive conclusions (IARC 1991). After evaluating scores of both animal and human studies, IARC rated chlorinated drinking water 'not classifiable' as to its carcinogenicity. Among the many exposure-related difficulties in making any connection between harm and consumption:

- Correlation studies are generally of uncertain validity. Exposure variables assessed for whole communities do not necessarily reflect exposure of individuals.
- In both correlation and case-control studies, information on the nature of the water source and chlorination status was obtained after or contemporaneously with the period over which cancer occurrence was measured. Because of the long latency between exposure and disease, it is better to correlate cancer rates with characteristics of water supplies identified before cancers occurred.
- Most studies didn't address the problem of migration in and out of studied communities over time.
- Recall bias was hard to remedy. Limited availability of water supply records hampered efforts to verify recollections.
- Water consumed outside the home, as well as the daily quantity of water consumed, were rarely taken into account.
- Exposure misclassification was common. Surrogates such as surface water, well depth and residence in community with chlorinated water supply can be used – but if these do not reflect exposure to chlorinated water during possibly relevant time periods for the aetiology of cancers in question, they will result in misclassification of subjects by exposure and will introduce bias.

As the IARC monograph noted, 'even in the best studies, errors in exposure measurement may still be a problem' (IARC 1991). Moreover, chlorinated water is different in different locations and cannot be considered to be the same entity. The relationship between chlorine dose and organic carbon present greatly affects the by-products formed; this in turn complicates the evaluation of whether chlorine residue maintained in chlorinated water or by-products in chlorinated waters or by-products of chlorination are responsible for any effect observed epidemiologically.

A more recent 'meta-analysis' of studies on chlorination, chlorination by-products and cancer 'suggest a positive association' between consumption of chlorination by-products in drinking water and bladder and rectal cancers in humans (Morris et al. 1992). The authors estimated that 9 per cent of bladder cancers and 18 per cent of rectal cancers per year are associated with the consumption of chlorinated water. However, the analysis acknowledges many of the same problems with exposure assessment noted by IARC – namely, inadequate information about the environment to which a subject was exposed (the source of tap water), the level of the proposed agent present in the environment (concentrations of chlorinated by-products in the tap water), and the degree to which the person was exposed to that environment (the amount of tap water consumed). In addition, the studies failed to adjust for diet as a confounding factor.

Finally, the authors emphasised that 'Our findings are in no way intended to suggest that the disinfection of drinking water should be abandoned. The potential health risks of microbial contamination of drinking water greatly exceed the risks described (in the study)...' It is an important caveat repeatedly issued by researchers – and repeatedly ignored by environmentalists and the media.

It's elementary

As Greenpeace activist, Christine Houghton, sees it: 'Since its creation, chlorine has been a chemical catastrophe ... It's either chlorine or us' (Greenpeace 1991). Yet chlorine is a ubiquitous element, one of the basic building blocks of all matter on the planet. (In its elemental state, chlorine is a greenish-yellow gas formed by passing electricity through salt water.) The chemical industry manufactures over 15 000 different chlorine-containing compounds; Mother Nature produces at least 1500 more (Gribble 1994a). In fact, scientists are only now beginning to discover and identify the great number of natural organohalogens present in our world.

Volcanic activity, forest and grass fires, fungi, algae, ferns and the decomposition of seaweed all release chlorinated organics into the environment (Willes et al. 1993). The smoke of burning wood alone contains more than 100 organochlorine compounds, including polychlorinated dioxins and polychlorinated dibenzofurans; the total annual global emission rate of chlorinated organic chemicals is 5 million tons (Gribble 1994b). By comparison, annual anthropogenic emissions total only 26 000 tons (Gribble 1994b). Our own bodies produce hypochlorite to fight infection and hydrochloric acid for proper digestion (Gribble 1995). And there is, of course, sodium chloride – common table salt – present naturally in mines, lakes and sea-water, found in our blood, sweat and tears, and essential to the diets of humans and animals (Winterton 1997). Even if Greenpeace's wish came true and all man-made sources of chlorine were shut down, natural mass production of compounds containing the condemned atom would continue undaunted. (See Table 7.1.)

The 'trouble' with chlorine is its sex appeal. Negatively-charged chloride ions have a high electron affinity that make them irresistibly attractive to other electron-rich atoms – most notably, carbon. Together, carbon and chlorine can be synthesised into a vast number of molecular structures called organochlorines. These and other chlorine-dependent processes and compounds, including polyurethanes, polycarbonates, epoxy resins, food wrap, insulation for electrical equipment, titanium dioxide (which whitens paint

Table 7.1: Chlorine in nature

Natural chlorinated organic chemical	*Source*
2,3,7,8 tetrachlordibenzo-p-dioxin	Forest and brush fires
2,4 dichlorophenol	Produced by lone star tick as sex pheromone
2,5 dichlorophenol	Secreted by grasshoppers
Methyl chloride	Marine algae, kelp, wood-rotting fungi
Hydrogen chloride	Volcanoes
Hypochlorite	White blood cells
Hydrochloric acid	Produced by stomach in humans to aid proper digestion

and toothpaste), silicones, and dry cleaning solvents, are integral to modern life. According to one industry-backed report, almost 85 per cent of the pharmaceuticals manufactured worldwide require chlorine at some stage of production (see Table 7.2); 96 per cent of crop-protection chemicals are chlorine-dependent; and up to 60 per cent of all chemical production depends on chlorine that can be added to, or removed from, other elements with ease and specificity (*The Lancet* 1992).

The long-lasting bonds that make some of these compounds so useful to humanity also make them baneful to environmentalists. As a writer for *Sierra* magazine explained it rather darkly, '[t]he hallmark of organochlorines is their tendency to bio-accumulate, and to pass from one generation to the next through the placenta. They are, in this respect, the molecular version of Original Sin' (*Sierra* 1995).

Sinful? Not exactly. It is true that chlorine's attractiveness to electrons tends to stabilise some chemical structures so that they don't break down easily. Some chlorinated chemicals persist for long periods until swept up into the stratosphere, where ultraviolet light helps break them into molecular fragments that destroy ozone. Others may persist in human fat tissue, where they may accumulate. As Willes et al. point out, however, the vast differences among chlorinated organic chemicals with respect to their physical and chemical properties and behaviour in the environment preclude the generalisation that all organic chemicals containing chlorine behave similarly in the environment and act as persistent, bio-accumulative chemicals (Willes et al. 1993).

Table 7.2: Uses of chlorine in pharmaceuticals

Chlorinated compound	*Pharmaceutical product*
Dimenhydrinate	Nausea preventatives
Pseudoehdrine hydrochloride	Decongestants
Procaine hydrochloride	Local anaesthetics
Chloroquine	Anti-malarial drug
Chlortetracycline	Antibiotics
Chlorpromazine	Anti-psychotics
Diphenhydramine hydrochloride	Antihistamine, cold and allergy treatments

Chemical convicts

Underlying the attack on chlorine is that since we 'know' that some of these chemicals, such as DDT and dioxin, are so terribly harmful to humans, we might as well ban the whole spectrum. As self-styled 'conservative environmentalist' Gordon Durnil, former chairman of the International Joint Commission, explains: 'we decided you can't distinguish among different compounds of chlorine as to which is harmful and which is not ... We decided we needed to look at chlorine as a class and decided because of the effects of dioxin, that use of chlorine as (an industrial) feedstock should be sunset' (Fumento 1994). But even if the cases against one or two chlorinated compounds were proven, it would hardly make sense to wipe out an entire element in the periodical table because of a few proven culprits with the huge spectrum known as organochlorines. 'Guilty-until-proven-innocent' has no place in the law or the courts – neither should it have one in science or public policy. The stakes for humanity are just too high.

A typical argument for class action against chlorinated chemicals, as illustrated by Greenpeace's Claire O'Grady Walsh, goes something like this:

'Seveso, Bhopal, the ozone hole, the greenhouse effect, Agent Orange, dioxin, DDT, PCBs have one thing in common – chlorine' (*The Irish Times* 1992). From there, the prohibitionists argue that 'trying to regulate thousands of organochlorine poisons one-by-one is doomed to failure. The phase-out strategy needs to be applied to the root cause – chlorine' (Rowley 1991). But as Phillipe Shubik, cancer researcher and toxicologist at Oxford University, observes: 'Any scientifically based toxicologist finds that kind of general approach abhorrent' (Amato 1993). Even environmentalist icon Mario Molina, an atmospheric chemist at the Massachusetts Institute of Technology, who first pointed out the link between chlorofluorocarbons and ozone depletion, has dismissed the movement to ban chemicals as a class: 'It isn't taken seriously from a scientific point of view', he told *Science* magazine (Amato 1993).

DDT

DDT (dichloro-diphenyl-trichloro-ethane) was essentially the first modern pesticide. It replaced highly toxic pesticides based on heavy metals, dramatically improved crop yields in America and abroad, and was instrumental in virtually wiping out malaria in the USA and many other countries by destroying the mosquito populations. Indeed, it has been credited with saving over 100 million lives worldwide.

The pesticide's fate was sealed when it became no longer a chemical but a symbol. Rachel Carson made it so in *Silent Spring*, although ironically she said that as bad as DDT was, other pesticides made it seem harmless in comparison. Neither did she implicate DDT as a carcinogen (Carson 1962).

Despite Carson's being demonstrably wrong on a number of important issues (for example, as discussed in Kemm, Chapter 1, she said that DDT threatened the robin with extinction, even as the increase in DDT usage coincided with a large increase in the robin population), establishing the lethality of DDT has become an environmentalist obsession. They got it banned, but had to cheat a bit to do it. Then-EPA Administrator William Ruckelshaus established the Agency's anti-science reputation in 1972 by overriding the decision of the hearing examiner who surveyed 9000 pages of documents from 125 expert witnesses on all sides of the issue. Among the examiner's findings, conclusions and orders: (Sweeney 1972).

- DDT is not a carcinogenic hazard to man
- The adverse effect on beneficial animals from the use of DDT under the registrations involved here is not unreasonable on balance with its benefit
- The Petitioners have not met fully their burden of proof
- There is a present need for the continued use of DDT for the essential uses defined in this case

In defiance of these recommendations, Ruckelshaus ordered a virtual ban on all uses of DDT effective January 1 1973. His Final Order stated boldly that '[T]he evidence on record showing storage [of DDT] in man and magnification in the food chain, is a warning to the prudent that man may be exposing himself to a substance that may ultimately have a serious effect on his health' (Ruckelshaus 1972). Yet his decision was based neither on existing research, nor on the examiner's findings – which Ruckelshaus himself failed to read.

With regard to the decline of raptor and pelican populations in the 1950s and 1960s (as well as the thinning of their eggshells), a majority of scientists agree that DDT contributed to these adverse effects. But in the opinion of the hearing examiner, these effects were 'not unreasonable on balance with' DDT's benefit. Since the curtailment of its use, declining tissue concentrations in wildlife species like the bald eagle and brown pelican have resulted in some degree of population recovery. These recoveries provide evidence, as Willes et al. point out, 'that any adverse effects of these chemicals are reversible and appear to be dose-related' (Willes et al. 1993). In other words, they refute environmentalists' claims that chlorinated organic chemicals cause irreversible effects and have no carcinogenic thresholds.

We know that whatever harm DDT may have caused, ceasing its use in many countries was absolutely catastrophic. Well-documented is the case of Sri Lanka, in which 2.8 million malaria cases per year in 1948 dwindled to but 17 cases after 15 years of DDT spraying. But after spraying was stopped in 1964, as a direct result of Carson's book, malaria cases quickly shot back up to their original level. Moreover, as DDT was phased out and alternatives, such as the organophosphate, parathon, phased in, mortalities increased significantly among farm workers. In addition to being more highly toxic in minute doses, these substitutes were more expensive and required more frequent applications than the vilified DDT.

Synthetic pyrethoids were developed over time to replace DDT as safe and effective alternatives, but they were not marketable in time to prevent the health and economic damage that the hasty DDT ban wrought. Now environmentalists are seeking a world-wide phase out of DDT. But as Salif Diop, a Senegalese delegate to the United Nations pointed out recently, 'In our countries we need chemicals like DDT to fight malaria. If you want a global ban, then you must come up with alternatives' (Chatterjee 1995). These substitutes must be safe, affordable and cost-effective – not merely chlorine-free – in order to do more good than harm.

PCBs

PCBs (polychlorinated biphenyls) were once widely used as liquid coolants, lubricants and insulators in industrial equipment, such as power transformers. Following a high-profile tragedy in

Kyushu, Japan, where over 1000 people contracted skin disorders from eating rice oil contaminated with high doses of PCB, researchers began to search for PCB residues in the environment. Despite lack of evidence at the time showing that trace amounts of PCBs in wildlife were causing harm, the mere presence and durability of the chemical provoked immediate regulatory activity.

At first, the Food and Drug Administration took sensible action – weighing both the health and economic costs and benefits of a ban, the agency decided instead to set practical tolerance levels of PCBs in fish. It noted in a standard-setting document that a complete ban on PCB residues would 'unnecessarily deprive the consumer of a portion of his food supply and disrupt the Nation's food distribution system' (US FDA 1973). Pressure from environmental groups for zero tolerance mounted, however, and in October 1976, President Gerald Ford signed the Toxic Substances Control Act – which required a complete phase out of all production and sales of PCBs – into law.

Evidence from animal studies was equivocal. Rats fed levels 5000 times the regulatory limit for humans developed excess liver tumours. Ignored was the fact that test rats actually had fewer reproductive-system cancers than expected, making their overall cancer rate no higher than that on untreated rats (Kimbrough et al. 1975). Reproductive effects of PCBs in high doses among monkey species were similarly mixed. Meanwhile, epidemiological studies among workers exposed to PCBs for prolonged periods have proved inconclusive. Studies conducted by the National Institute for Occupational Safety and Health on workers who inhaled or absorbed PCBs through their skin over many years have shown no adverse human health effects (Smith 1981). One recent study even showed slightly lower rates of cancer deaths and deaths in general than expected (Brown 1987). Nevertheless, the EPA continues to require the diversion of billions of dollars to eradicate virtually all traces of PCB residues in soil under the conservative assumption that some amount somewhere might potentially cause harm to a dirt-digesting toddler who, even if exposed to 10 ppm per day, would absorb 125 000 times less than the amount in the daily diet associated with increased cancer incidence among two strains of rats.

Dioxin

For 15 years now, the EPA has clamped down on dioxin, a by-product of paper bleaching and of incineration of certain materials. Until it was banned as such, it was also a by-product in the manufacturing of some herbicides, including the notorious Agent Orange. Ever since the chemical was found to be horribly toxic to guinea pigs – albeit far less so to every other animal species tested, including other rodents – the EPA and other environmental organisations have relentlessly attacked it as they have no other chemical except for the pesticide DDT. Fear of dioxin contamination led to the evacuation of Love Canal near Niagara Falls in 1978 and of Times Beach, Missouri in 1983, and to telling veteran soldiers of the Vietnam War that they may be at extraordinary risk of disease.

But while dioxin was long touted as 'the most deadly chemical created by man', decades of scientific scrutiny have found that its only acute human effect is a form of acne (Neubert 1997). As dioxin expert Dr Michael Gough noted:

> No human illness, other than the skin disease chloracne, which has occurred in highly exposed people, has been convincingly associated with dioxin. In short, epidemiologic studies in which dioxin exposures are known to have been high, either because of the appearance of chloracne or from measurement of dioxin in exposed people, have failed to reveal any consistent excess of cancer. In those studies that have reported associations between exposure and disease, no chloracne was reported, and there are no measurements of higher-than-background levels of dioxin in the people who are classified as exposed (Gough 1991).

The case against dioxin for threatening people is almost as suspect as that against DDT, but this hasn't deterred environmentalists. For years they charged that dioxin was a powerful human carcinogen but epidemiological studies failed to back them up. So, quietly, they began to shift the accusation from the 'most deadly' carcinogen known to man, to the most potentially damaging to unborn children, immune systems and hormones.

A recent dioxin assessment report by the EPA made the shift official. In addition to the old charge of dioxin being a possible human carcinogen – causing as many as one in 1000 human cancers – the EPA added two newer charges: that it might affect human children in the womb and that it could compromise immune systems at levels approaching those to which Americans are currently exposed. While those human exposure levels are infinitesimal compared with our exposure to many other chemicals, the EPA maintained that what causes illness in some animals at huge doses must also cause sickness in humans in tiny ones.

But several scientists at the meeting challenged the EPA's assumption, used in all its policy-making, that there is not a threshold below which a harmful chemical causes no harm. One was the University of Wisconsin's Alan Poland, widely known for his discovery of the 'dioxin receptor', the molecule in cells to which dioxin must bind before it produces any affects. He said that 150 years of science contradicted the EPA no-threshold position.

Regarding dioxin, Dr Poland said that the normal level to which Americans are exposed – four molecules of dioxin per cell – is far below the number required to have an effect, considering that there are about 10 000 receptor molecules per cell. One EPA official complained that its science advisory board (SAB) meeting had unfairly been characterised as negative. In fact, he said, the only problems the SAB found were in the ninth chapter. Of the first eight, he was highly complimentary

But that's just the point. The first eight chapters were written by scientists outside the EPA. Only the last chapter, the conclusive one, the one from which EPA was to draw its regulatory policy and from which the media drew the headlines, was the one written by the EPA itself. In that chapter, said Dr Poland at the meeting, 'policy masquerades as science'. 'This is probably the best data that the EPA will see in my lifetime,' Dr Poland added, 'yet, despite all of that, the first eight chapters are thrown away' (*Food Chemical News* 1995).

Advisory board members repeatedly accused the EPA of picking and choosing its data. For example, the largest, most heavily-studied group of persons with known high exposure to dioxin were the members of Operation Ranch Hand, the men who did the actual spraying of Agent Orange on the jungles of South Vietnam. The EPA report duly noted any possible minor abnormality in this group. But it neglected to say the Ranch Handlers were strapping specimens of healthy humanity. 'The EPA didn't mention that there were no more cancers than would be expected, no effects on the immune and nervous systems, no increase in deaths, and no increased birth defects in their children' SAB member and Office of Technology Assessment official, Michael Gough noted. 'They mentioned nothing that didn't serve their purpose' (*Food Chemical News* 1995).

Also unmentioned were follow-up studies conducted in Seveso, Italy, where 37 000 people were exposed to high doses of dioxin following the explosion of an unattended chemical reactor. As the Institute of Occupational Health at the University of Milan found, there were 'no increased birth defects due to dioxin exposure' (Bertazzi 1991). Furthermore, cancer mortality rates were inconclusive.

One SAB member, Dr Knute Ringen of the Center to Protect Workers' Rights in Washington DC, concluded: 'I think that the agency has pretty much come to the end of the line with regard to producing useful decision-making information on dioxin, and that it's time to go on to something else' (Bertazzi 1991, Gribble 1995, WHO/IPCS 1998).

Gender wars

The assault on chlorine has recently focused on its alleged gender-bending characteristics. Chlorinated organic compounds bind to oestrogen receptors in cells, the theory goes, which may lead to changes in those cells, tissues or organs.

Environmentalists blame process-related organochlorines such as dioxin for altering the sexual characteristics of fish. In particular, chlorinated dioxins and furans from pulp mill effluent have been identified as culprits in causing sex changes among fish (Rogers 1989). As Willes et al. (1993) note, however, there is increasing evidence that something other than the usual chemical suspects may be largely responsible for the effects noted – namely, natural plant sterols (phytosterols), which were implicated as a causal factor in the masculinised female mosquito-fish observed in a north-west Florida stream receiving pulp and paper mill effluent.

Environmental oestrogens have also been accused of causing breast cancer in women. In 1993, a highly publicised report in the *Journal of the National Cancer Institute* found that DDE (a metabolite of the oestrogenic pesticide DDT) was present in higher concentrations among a small population of Long Island breast cancer victims when compared to a control group (Wolff et al.

1993). Concentrations of DDE were about 35 per cent higher in the women with cancer than in the controls.

It is possible that high blood levels of DDE increase the risk of breast cancer, but the link probably runs in the opposite direction. Chronic diseases like breast cancer mobilise fat from fat storage deposits, which contain fat-soluble compounds like DDE. As a result, blood concentration levels of DDE increase. Moreover, certain drugs can also increase blood levels of DDE. In short, despite the media's conventional wisdom, the 1993 *JNCI* study does not prove a causal relationship between DDE and breast cancer (Wolff et al. 1993).

A larger study also published in *JNCI* reported no association between breast cancer and higher levels of either DDE or PCBs (Krieger et al. 1994). As toxicologist Michael Gallo of the Robert Wood Johnson Medical Center at Rutgers University noted: 'The scare was that these oestrogens were so potent that they were causing an increased incidence in breast cancer. This latest study quieted down those fears' (Stone 1994).

While several other publications have reported increased concentration of bio-accumulative organochlorines in human tissues, there is not consistency among these various studies in the association of the increased tissue concentrations and specific human diseases (Willes et al. 1993). Nor have increased mammary tumours been shown in laboratory studies, where doses and confounding factors are much easier to control in comparison to epidemiological studies. In short, the balance of evidence does not support a causal association between exposure to organochlorines and increased human breast cancer incidence.

In fact, women produce far more of their own oestrogen than they could ever possibly absorb from the environment. Background levels of synthetic oestrogens are dwarfed by the body's own production of oestradiol. To put it in proper perspective, toxicologist Stephen Safe of Texas A&M University notes that 'The average human exposure to oestrogens is 99.999 per cent from natural sources' such as fruits and vegetables (Graff 1995). Even giving women extra doses of their own oestrogen, either as post-menopausal hormone therapy or as birth-control pills, increased cancer risk either slightly or not at all (see Chapter 9).

Safe also points out that some organochlorines such as 2,3,7,8-tetrachlorodibenzo-p-dioxin (TCDD), actually exhibit anti-oestrogen activity that may counteract adverse health effects including breast cancer. As he reported at the 1994 annual meeting of the Society of Toxicology, data from several studies of a protein in cells found in breast cancer tumours show that TCDD can block production of the targeted protein. 'Dioxin in combination with oestrogen block all of these (oestrogen) responses', Safe told the audience (*BNA Chemical Regulation Daily* 1994). Other organochlorines such as DDT and PCBs may also act as anti-oestrogens.

The mismeasure of man

Perhaps the most potent weapon in the anti-chlorine activists' arsenal has been the charge of falling sperm counts. The alarm over chemically-induced male infertility was prompted by a 1992 study from Copenhagen that claimed to show that sperm concentration per unit volume had fallen by over 40 per cent between 1940 and 1990 (Carlsen 1992). A year later, one of the authors of the study, Niels Skakkebaek, penned a follow-up piece with Richard Sharpe of the British Medical Research Council Reproductive Biology Unit in Edinburgh, Scotland, in *The Lancet*. The duo speculated that foetal exposure to synthetic oestrogens may be the prime suspect in the sperm count crisis (*The Lancet* 1993 and see Chapter 10).

Unsurprisingly, *The Lancet* article prompted a barrage of alarmist media reports. The BBC aired a documentary entitled 'The Assault of the Male'; in the US, Connie Chung devoted her now-defunct investigative show, 'Eye to Eye' to the plight of infertile men under chemical siege. As scientists have noted, however, the original study – a 'meta-analysis' of 61 studies on falling sperm counts – suffered from numerous statistical and methodological shortcomings.

First, the Danish researchers included studies irrespective of their sample size, many of which were so small that they would not normally be considered admissible evidence (Farrow 1994). In an editorial for the *British Medical Journal*, Stephen Farrow of Middlesex University's Health Research Centre noted that one study was of seven men; 11 others were of fewer than 20 men; and another 29 were of fewer than 50 men (Farrow 1994). These were 'given greater weight than they deserved', Farrow wrote, through the misapplication of statistical tests (Farrow 1994).

After investigating the Skakkebaek data, a different team of researchers showed that nearly all of the alleged decreases could be explained by the changing definition of a normal sperm count over the past 50 years (Bromwich et al. 1994). Bromwich et al. concluded in a study also published by the *BMJ* that 'The original evidence does not support the hypothesis that the sperm count declined significantly between 1940 and 1990' (Bromwich et al. 1994).

Furthermore, critics who reanalysed the data have challenged the timing of the decline reported. As reproductive specialists Anna Brake and Walter Krause of Philipps University in Marburg, Germany explained in a letter to the *BMJ*, 48 of the studies used in the meta-analysis were published since 1970 – accounting for 88 per cent of the men studied (Brake and Krause 1992). These studies actually showed a slight increase in sperm counts. Krause and Brake concluded that 'care should be taken when discussing a causal relation with environmental factors' (Brake and Krause 1992).

The plea for caution fell upon deaf ears at Greenpeace headquarters. Seizing on the Skakkebaek findings, the environmental group launched a new advertising campaign that publicised the alleged environmental threat to man's virility. 'You're not half the man your father was', the ads taunted.

Responsible scientists assailed environmentalists for exploiting the data. As one set of outraged researchers argued, 'there is no conclusive evidence' for blaming exposure to oestrogenic chemicals in the environment for falling sperm counts or shrinking penises (Connor 1995). The hypothesis 'is based on evidence too limited to allow firm conclusions to be drawn. It is premature to call for a ban on these or any other chemicals before more research is done. They are misrepresenting this research... They are taking something which is a clearly stated hypothetical link and calling it fact' (Connor 1995). The identity of these critics? None other than Niels Skakkebaek and Richard Sharpe, whose research prompted the sperm scare in the first place.

Rachel's folly

In *Silent Spring*, Rachel Carson was the first to bear witness against chlorinated hydrocarbons. But there is nothing magical nor artificial about attaching a chlorine atom to a carbon atom, which is all a chlorinated hydrocarbon is. Some chlorine-based compounds like DDT may persist in body fat year after year; some do not. Some cause cancer in laboratory animals fed massive doses; others do not.

Chlorinated compounds comprise all major chemical classifications, including inorganic salts and acids, aromatics and aliphatics. As Willes et al. point out, 'the presence of a significant degree of chlorination is not, of itself, sufficient to confer bio-accumulative potential on a compound' (Willes et al. 1993). PCBs, chlorinated dioxins, chloroform, and trichlorophenol all may share a chlorine link, but vast differences in their physical and chemical properties lead to vastly different behaviour in the environment (Willes et al. 1993). Depending on the position and number of substitutions, adding a specific chlorine group to an organic molecule may increase or reduce the chemical reactivity of that molecule. Thus, the attempt to eliminate arbitrarily an entire class of chemical compounds based on a few 'bad' molecules is sweepingly over-broad.

The high price of 'precaution'

Chlorine's critics argue that we can't wait for the scientific evidence to roll in. Demonstrating this 'ban now, ask questions later' approach, a 1993 Greenpeace report attempting to link chlorine and breast cancer declared: 'If proof is defined as evidence, beyond any doubt, of a cause-effect link between individual chemicals and the disease, in which all confounding influences have been eliminated, the answer is no…' (Thornton 1993). But, it went on, 'It is unethical, irresponsible, and unrealistic to require strict proof, because such an approach takes preventative action only after irreversible damage to health and environment have taken place.'

Or as Lois Gibbs, the homeowner whose alarmist calls led to the unjustified evacuation of Love Canal, New York, writes in her recently published book, *Dying from Dioxin: A Citizen's Guide to Reclaiming Our Health and Rebuilding Democracy*: Government should make environmental regulatory decisions 'so that the burden of proof is placed on showing that a chemical or practise is safe, not is harmful … Rather than using all chemicals until they are proven harmful, we should demand that all chemicals be shown to be safe before they are used' (Gibbs 1995).

The so-called precautionary principle lends an almost irresistible sense of moral urgency to the environmental movement. Unless such a policy is adopted, the argument goes, we will be saddled with policies that wait for a 'body count' before prudent action is taken. Those who favour the principle favour health; those who oppose it oppose saving our children and our environment. But at what price prudence?

An industry-sponsored study concluded that 'about 45 per cent of all US industries are direct consumers of chlorine and its co-products, and that all industries are indirect consumers of chlorine or chlorine-dependent products' (Charles River Ass. 1993). The researchers also estimated the total net cost of substitution to consumers in the United States would be slightly over $90 billion, and that employment in chlorine-dependent industries accounted for £33.6 billion in wages in 1990.

The price of precaution can not only be measured in dollars, but in lives. As risk analyst Ralph Keeney has shown, mortality risks induced by economic expenditures are significant. He has estimated that every $7.25 million taken out of the economy by government results in the loss of one human life on average. Thus, anti-chemical campaigns that do not take into account the possibility of risk or death associated with implementing bans are of little value to public health at all. As Keeney notes. 'if the intent of a proposed regulation is to save lives by making some aspect of life safe, then it would seem ridiculous not to consider the potential mortality implications of implementing the regulation itself. These implications include the potential fatalities induced by the cost of the regulation' (Keeney 1990).

The precautionary principle is a something-for-nothing proposition. Environmentalists would have us believe that improved health can be attained through regulations and bans at zero cost. They assume costless transitions to chlorine-free substitutes. But government intervention requires resources – resources diverted from other proven health-improving and cost-effective activities. Furthermore, cost-effective alternatives aren't free. The ban on DDT led to increased malarial infections and deaths. Alarm over chlorinated water contributed to the deadly Latin American cholera epidemic. The phase-out of chloroflourocarbons (CFCs), another chlorine-containing culprit, led to increased water pollution by the electronics industry.

Chlorine processes and products didn't simply arise out of thin air to wreak havoc on the environment; they developed over time to replace older, more hazardous technologies. Before the successful widespread introduction of chlorination to purify water, for example, treatment techniques included filtration, followed by chemical precipitation and sedimentation methods (IARC 1991). These methods, however, could not guarantee a bacteriologically safe water supply (IARC 1991). Chlorine is by far the superior method of disinfection because it is effective against a broad spectrum of pathogens including bacteria, viruses and protozoa; only state-of-the-art chlorine chemistry provides residual protection, that is, the ability to prevent microbial growth after treated water enters the distribution system; and finally, chlorine disinfection technology is far simpler than other disinfection technology (AWWA 1995).

In most cases, the proposed 'cure' for chlorinated organic chemicals is worse than the 'disease'. Environmentalists claim, for example, that substitutes for water chlorination are cheaper and safer. In fact, they may prove far more odious. Ozone decomposes too quickly to provide any residual disinfection, against recontamination of water supplies. Chlorine or chloramines must be added to provide residual protection; thus, ozone could not serve as an adequate substitute for chlorine on its own. Moreover, ozone must be generated on site, is not as suitable as chlorine for smaller treatment works, and also results in various mutagenic by-products. As a team of international researchers concluded in *JAMA*, there simply are 'no cheap substitutes as proven and effective as chlorination' (Mintz et al. 1995).

Conclusion

The proper way of discriminating among chemicals isn't between chlorines and non-chlorines, or between naturals and synthetics. Rather, our goal should be to use quality science on a case-by-case basis to restrict and seek alternatives to the usage of any chemical, chlorinated or not, synthetic or not, which causes demonstrable harm to humans or their environment. This won't accomplish any political, moral, religious goals. But it will make the world a safer place.

Ultimately, however, the war over chlorine and all its diverse compounds, both naturally-occurring and synthetically induced, will not be won with the weapon of 'sound science.' The conflict is not between those who desire better health and safety and those who do not. It is between those who believe increased wealth and technological progress are the best means of improving health and safety and those who do not. Environmentalists see rapid technological advancement as a threat to human dignity. 'Sound science' will not thwart their campaign because they are suspicious of science itself. As long as the environmentalists' view that all synthetic chemicals are 'guilty until proven innocent' prevails, no amount of exculpatory evidence will ever be enough to clear an arbitrarily indicted chemical.

As the anti-chlorine activists themselves have framed the debate, 'It is either chlorine or us'. Chlorine-based products and processes are essential to modern life. Technology-fearing environmentalists are not. The choice should be clear.

References

Amato, I. (1993). 'The Crusade against Chlorine,' *Science*, **261**, 5118:152–154.

American Public Health Association (1993). *Recognizing and Addressing the Environmental and Occupational Health Problems Posed by Chlorinated Organic Chemicals*, APHA Policy Statement 9304 1948–present, cumulative (Washington, DC: American Public Health Association), pp 519–520.

American Water Works Association (AWWA) (1995). 'Chlorine for Drinking Water Disinfection', white paper, *AWWA Mainstream* (American Water Works Association Washington, DC).

Anderson, C. (1991). 'Cholera epidemic traced to risk miscalculation', *Nature*, **354**, November 28, 255.

Bertazzi, P. A. (1991). 'Long-term effects of chemical disasters: lessons and results from Seveso', *Science of the Total Environment*, 106:5.

Brake, A. and Krause, W. (1992). Letter to the *British Medical Journal*, December 12.

Bromwich, P. et. al. (1994).'Decline in sperm counts: an artefact of

changed reference range of "normal?"', *British Medical Journal*, **309**, 6946:19, July 2.
Brown, D. P. (1987). 'Mortality of Workers Exposed to Polychlorinated Biphenyls – An Update', *Archives of Environmental Health*, **42**:333-339.
Bureau of National Affairs (BNA) (1993). 'Organic Compounds Pose Risks to Public, Health Association Says', *BNA Chemical Regulation Daily* Nov, 1.
Bureau of National Affairs (1994). 'Detailed administration proposals on clean water rewrite released by EPA', Daily Report for Executives, February 2, Section A, p. 21.
BNA Chemical Regulation Daily (1994). Quoted in 'Anti-estrogen benefits of TDCC overlooked; chemical may be treatment for breast cancer', March 18.
Carlsen, E. et. al. (1992). 'Evidence for the decreasing quality of semen during the past 50 years', *British Medical Journal*, **305**:609–12.
Carson, R. (1962). *Silent Spring*. New York: Houghton Mifflin.
Charles River Associates Incorporated (1993). Assessment of the Economic Benefits of Chlor-Alkali Chemicals to the United States and Canadian Economies, April. (Washington DC).
Chatterjee, P. (1995). 'Land-Based Toxic Chemicals Killing Ocean Life', November 2. (United Nations, New York).
Connor, S. (1995). '"Green" evidence of smaller penises does not stand up', *The Independent*, June 7, p. 6.
Cox, J. R. (1994). Sierra Club President, quoted in *The Planet*, (Washington, DC: Sierra Club), November.
Dunnick, J. K. and Melnick, R. L. (1993). 'Assessment of Carcinogenic Potential of Chlorinated Water: Experimental Studies of Chlorine, Chloramine, and Trihalomethanes', *Journal of the National Cancer Institute* **85**, 10, May 19.
Durnil, G. (1995). *The Making of a Conservative Environmentalist*, Bloomington, Indiana: Indiana University Press.
Farrow, S. (1994). 'Falling sperm quality: fact or fiction?' Editorial, *British Medical Journal*, **309**,6946:1, July 2.
Food Chemical News (1995). 'EPA Dioxin Assessment slammed for lack of science basis', September 11, 37, 29.
Fumento, M. (1994). 'Chemical warfare: campaign to ban chlorine', *Reason*, **26**,2:42 June.
Gibbs, L. (1995). *Dying From Dioxin: A Citizens Guide to Reclaiming Our Health and Rebuilding Democracy*, South End Press, Washington D C.

Gough, M. (1991). 'Human health effects: what the data indicate', *Science of the Total Environment* 104:129. Greenpeace press release, July 26 1991.
Graff, G. (1995). 'The chlorine controversy', *Technology Review*, January, **98**, 1:54.
Gribble, G.W. (1994a). 'Natural organohalogens', *Journal of Chemical Education*, **71**:907–911.
Gribble, G. W. (1994b). *Environmental Science and Technology*, 28, 310A 1994, cited in: Abelson, Philip H., 'Chlorine and organochlorine compounds' Editorial, *Science*, **265**, 5176:1155 August 26.
Gribble, G. W. (1995). Chlorine and Health (New York: American Council on Science and Health, August).
Hilleman, B. (1993). 'Concerns Broaden over Chlorine and Chlorinated Hydrocarbons', *Chemical and Engineering News*, April 19, p. 11.
International Agency for Research on Cancer (IARC) (1991). 'Chlorinated Drinking Water; Chlorination By-products; Some other Halogenated Compounds; Cobalt and Cobalt Compounds', *IARC Monographs on the Evaluation of Carcinogenic Risks to Humans*, 52: 45–141. (Lyon, France: International Agency for Research on Cancer).
International Joint Commission (IJC) (1990). Fifth Biennial Report on Great Lakes Water Quality, Part II (Ottawa–Washington: International Joint Commission) p. 7.
International Joint Commission (IJC) (1992). Sixth Biennial Report on Great Lakes Water Quality (Ottawa–Washington: International Joint Commission) pp. 28–30.
International Joint Commission (IJC) (1994). Seventh Biennial Report on Great Lakes Water Quality (Ottawa–Washington: International Joint Commission).
The Irish Times (1992). 'Greenpeace blocks ICI chlorine ship', July 6.
Keeney, R. (1990). 'Mortality Risks Induced by Economic Expenditures', *Risk Analysis*, **10**, 1:157.
Kimbrough, R. D. et al. (1975). 'Induction of Liver Tumors in Sherman Strain Female Rats by Polychlorinated Biphenyl Aroclor 1260', *Journal of the National Cancer Institute*, **55**:1453-1456.
Krieger, N. et al. (1994). 'Breast cancer and serum organochlorines: A perspective study among white, black and Asian women', *Journal of the National Cancer Institute*, **86**:589–599.
The Lancet (1991). 'Cholera in Peru', **337**, March 2.
The Lancet (1992). 'Of cabbages and chlorine: cholera in Peru', **340**, July 4.

Larson, J. L., Wolf, D. C. and Butterworth, B. E. (1994). Fund Appl. *Toxicol.*, **22**, 90.
'Letter to the President on his recycled paper purchasing order signed by fourteen major environmental groups', September 23 1993.
McCarty, L. S. (1994). 'Chlorine and organochlorines in the environment: a perspective', *Canadian Chemical News*, **46**, 3, March.
Mintz, E. D. et al., (1995). 'Safe water treatment and storage in the home: a practical new strategy to prevent waterborne disease', *Journal of the American Medical Association*, **273**, 12, March 22
Moberg, D. (1993). 'Sunset for Chlorine?', *F Magazine*, July/August, **IV**,4.
Morris, R. D. et al. (1992). 'Chlorination, Chlorination By-products and Cancer: A Meta-analysis', *American Journal of Public Health*, **82**, 7:955–963, July.
Neubert, D. (1997). *Teratogenesis, Carcinogenesis and Mutagenesis*, 17:157–215, on toxicity of dioxin to humans.
Rogers, I. H. et. al. (1989). 'Observations on overwintering juvenile chinook salmon exposed to bleached kraft mill effluent in the upper Fraser River, British Columbia', *Chemosphere* **19**, 12:1853–1868.
Rowley, S. R. (1991). 'Greenpeace: Chlorine Polluting Great Lakes', *Chicago Tribune*, July 26.
Ruckelshaus, W. (1972). 'Consolidated DDT Hearings: Opinion and Order of the Administrator', *Federal Register*, **37**, July 7.
Sierra (1995). 'Elemental enemy: environmental hazards of chlorinated chemicals', **80**, 1:30.C January.
Sharpe, R. M. and Skakkebaek, N. E. (1993). 'Are oestrogens involved in falling sperm counts and disorders of the male reproductive tract? Hypothesis', *The Lancet*, **341**, 8857:1392, May 29.
Smith, A. B. (1981). Cross section survey of two groups of workers occupationally exposed to PCBs in the maintenance, repair and overhaul of electrical transformers (Cincinnati, OH: National Institute for Occupational Safety and Health).
Stone, R. (1993). 'Environmental estrogens stir debate' Quotation. *Science*, **265**, 1570:303, July 15.
Stone, R. (1995). 'A molecular approach to cancer risk: in a major change in how toxic substances are tested and regulated, federal agencies will soon require molecular data on how chemicals cause cancer', *Science*, **268**, 5209:356, April 21.
Swerdlow, D. L., Mintz, E. D., Rodriguez, M. et al. (1992). 'Waterborne transmission of epidemic cholera in Trujillo, Peru:

lessons for a continent at risk', *Lancet*, **340**. July 4.
Sweeney, E. M. (1972). Hearing Examiner's Recommended Findings, Conclusions, and Orders, consolidated DDT hearings, (Washington, DC: Environmental Protection Agency, Apr. 25), p. 93.
Thornton, J. (1991). *The Product is the Poison: The Case for a Chlorine Phase-Out*, Greenpeace, Washington, DC.
Thornton, J. (1993). *Chlorine, Human Health and the Environment: The Breast Cancer Warning*, Greenpeace, Washington, DC.
US Congressional Press Releases, April 6 1995.
US Environmental Protection Agency (EPA) (1979). *National Interim Primary Drinking Water Regulations, Control of Trihalomethanes in Drinking Water,* Part III, Environmental Protection Agency, Final Rule, *Federal Register* 44: 68624–68707, November 29.
US Environmental Protection Agency (EPA) (1988). *Comparative Health Effects Assessment of Drinking Water Treatment Technologies*, EPA Publ. No. 570/988–009 (Washington, DC: EPA).
US Environmental Protection Agency (EPA) (1994). President Clinton's Clean Water Initiative, February.
US Food and Drug Administration (FDA) (1973). 'Polychlorinated Biphenyls. Contamination of Animal Feeds, Foods, and Food-Packaging Materials', *Federal Register*, 38: 18096–18102 July 6.
US House of Representatives (1993). H.R. 2898, p.6, lines 6–10.
Weekly Epidemiol Rec. (1993). World Health Organization, 'Cholera update, end of 1993,' 69:13–20.
Wolff, J. et al. (1993). 'Blood levels of organochlorine residues and risk of breast cancer', *Journal of the National Cancer Institute*, **85**: 648–652.
WHO (1998). *World Health Report* and see www.who.org
WHO/IPCS (1998). *Review of TDI for dioxins* (May – press release).
Wildavsky, A. (1988). *Searching for Safety*, New Brunswick, NJ: Transaction Publishers.
Willes, R. F., Nestmann, E. R., Miller, P. A. et al. (1993). 'Scientific Principles for Evaluating the Potential for Adverse Effects from Chlorinated Organic Chemicals in the Environment', *Regulatory Toxicology and Pharmacology*, **18**:313-356.
Winterton, N. (1997). *Mutation Research*, **373**, 2:293–294.

8 Organochlorines – natural and anthropogenic

Gordon W. Gribble

Of the 3100 known naturally occurring organohalogen compounds, more than 1800 contain chlorine. These far outnumber industrial organochlorines in use. Natural organochlorines, which range in structural intricacy from the simple ubiquitous fungal and plant metabolite chloromethane to the complex life-saving antibiotic, vancomycin, are produced by marine and terrestrial plants, bacteria, fungi, lichens, insects, marine animals (sponges, sea hares, nudibranchs, gorgonians, tunicates), some higher animals, and a few mammals including humans. New examples are continually being discovered and the total number of natural organohalogens may surpass 3500 by the turn of the century.

Given the campaign to ban summarily chlorine and industrial chlorination processes by several environmental groups – a crusade that has reached worldwide proportions – it is important to hear another side of the chlorine story.

Forty years ago, the few known naturally-occurring organochlorine compounds were considered chemical freaks, not to be taken seriously. In 1968, a scientist wrote: 'present information suggests that organic compounds containing covalently bound halogens are found only infrequently in living organisms' (Fowden, 1968). However, since then the number of natural organochlorine compounds has grown to more than 1800, most of which are produced by living organisms, such as marine and terrestrial plants, bacteria, fungi, lichen, insects, marine animals, and some mammals. Others are produced by natural thermal processes. Some of these chemicals are identical to man-made organochlorines with which we are familiar: chlorophenols, chloroform, dioxins, and CFCs. But many others are entirely new molecular entities that possess extraordinary biological properties similar to those of penicillin, morphine, and the new anti-cancer drug, taxol. New organochlorine compounds are being discovered constantly, as chemists search the biosphere for new medicinal drugs. This extraordinary explosion of information was the subject of the first 'International Conference on Naturally Produced Organohalogens' in The Netherlands in 1993.

As a fundamental chemical element, chlorine is not only abundant in the earth's crust (ranking eighteenth in the list of elements) but is also ubiquitous in our soil, rivers, lakes, trees, plants, and, of course, oceans. Like the other common elements that are present in all living things – carbon, hydrogen, oxygen, nitrogen, sulphur, phosphorus – so, too, is chlorine and the other halogens (bromine, iodine, and, to a lesser extent, fluorine). The major source of natural hydrogen chloride, which is the gaseous form of hydrochloric or stomach acid, in the atmosphere is volcanoes, and several recent eruptions have been studied with regard to their gas emissions (Symonds et al., 1988). For example, the Mt. Pinatubo eruption in 1991 released 4.5×10^6 tons of chlorine into the atmosphere, and the 1976 Mt. Augustine eruption in Alaska emitted 0.6×10^6 tons of chlorine. Likewise, Hekla in 1970, El Chichón in 1982, several Guatemalan volcanoes in 1978, and Hawaii's Kilauea eruptions all produced large quantities of chlorine, mainly as hydrogen chloride. Hydrogen chloride has also been detected outside our solar system. Both Mt. St. Helens and Kilauea, the latter of which has been erupting continually since 1983, also produce the simplest organochlorine compound, chloromethane (CH_3Cl). The Kamchatka volcanoes in Siberia and the Santiaguito volcano of Guatemala emit CFCs (Freons) in quantities well above background levels. Volcanic emissions include tetrachloroethylene, chloroform, carbon tetrachloride, methylene chloride, and several of the CFCs, chemicals formerly thought only to result from the actions of man. From one Kamchatka volcanic vent, CFCs were detected in levels of 160 ppb, which is 400 times that of the background atmosphere (Isidorov, 1990). Such natural chlorine and fluorine compounds are produced in the high temperature eruption zone of the volcano by the combustion of organic material, such as vegetation, sediments, or fossil soils in the presence of chloride and fluoride mineral deposits.

Organochlorine compounds are produced naturally by countless marine creatures (sponges, corals, sea slugs, tunicates, sea fans, jelly fish) and seaweeds, plants, seeds, trees, fungi, lichen, algae, bacteria, microbes, insects, and even man. Nearly 40 different organochlorine compounds are present in 'limu kohu', the favourite edible seaweed of most Hawaiians. Other seaweeds produce chloroform, carbon tetrachloride, and even the dry cleaning solvents, perchloroethylene and trichloroethylene, are produced by more than 20 species of marine algae (Abrahamssen et al., 1995). The 'smell of the ocean' is probably due to these volatile organochlorines and other organohalogens, which seem to serve as antibacterial agents for these seaweeds.

Many organochlorine compounds are used by the organism as hormones, repellents, pesticides, or to facilitate food gathering. Vegetables such as lentil, green peas, fava bean, vetch, and sweet pea produce the organochlorine 4-Cl-IAA as a growth hormone. The German cockroach manufactures two chlorinated steroids as pheromones, a dozen species of tick produce 2,6-dichlorophenol as a sex attractant, and locusts synthesise chlorinated tyrosines to strengthen the cuticle. A **Penicillium** mould produces 2,4-dichlorophenol, the same chemical which is one of the two herbicides in Agent Orange! Recently, an Ecuadorian frog was found to produce a chlorinated alkaloid, epibatidine, which is 500 times more powerful than morphine as a pain-killer. Perhaps the most medicinally important organochlorine compound is the microbial product vancomycin, the antibiotic and the drug of choice to treat penicillin-resistant **Staphylococcus** infections – the lethal 'super germ' – particularly those infections that occur in hospital patients. Yearly sales of the life-saving vancomycin and the related organochlorine natural antibiotic teicoplanin are in the hundreds of millions of dollars.

Research has shown that these natural organochlorine compounds are not derived from pollutants but are biosynthesised by individual organisms for very specific purposes. They are not, as some would have us believe, a product of the chemical revolution. As do other natural chemicals, these organochlorines play an essential role in the survival of the organism, and the ability to synthesise these compounds has evolved over aeons in the process of natural selection.

A remarkable recent discovery is that several chlorinated compounds have been isolated from humans (Winterton, 1997; Hazen and Heineche, 1997). These chlorine-containing amino acids and cholesterol derivatives are produced during the action of chlorine on invading bacteria and other pathogens. Indeed, chlorine (bleach) is actually produced by our white blood cells as part of our natural immune system! Thus, numerous studies have shown that cellular peroxidase enzymes in mammalian white blood cells transform natural blood chloride into free chlorine, resulting in the death of the pathogen (bacteria, yeast, fungi, and even tumour cells) by chlorination. This process is exactly analogous to the disinfection of drinking water with chlorine. This natural chlorination process may be the explanation for the higher concentration of organochlorines in human urine than in drinking water. It seems likely that many more organochlorine by-products produced during this 'biodisinfection' will soon be isolated and identified, and this research should provide insight into the human immune process. Biochlorination is apparently as natural a biological process as blood clotting or salivation.

Nature is an amazing recycler of chemicals. Insects, micro-organisms, and fungi all conspire to break down dead matter to simpler chemicals for reuse in the natural scheme of things. Many species of white-rot fungi decompose dead trees and other forest plant material in a process akin to what occurs in a garden compost pile. One of the by-products of this natural decay process is chloromethane, and 5 million tons per year of this chemical are produced naturally, dwarfing the 26 thousand tons produced annually by man. It may be that this natural chloromethane is an innate regulator of the stratospheric ozone layer. At least 25 per cent of the chlorine in the stratosphere is naturally produced chloromethane. Chloromethane is also produced in forest fires and is a biochemical regulator in many species of fungi, algae, evergreen trees, mushrooms, cabbage, and potato. Because fires have occurred on earth since land plants evolved 350–400 million years ago, it is clear that chloromethane has been a natural component of our atmosphere and stratosphere for aeons. The massive Australian brush fires of a few years ago produced large quantities of chloromethane. Chloromethane has been detected in mines and mineral processing plants, and is present in several ores and minerals (Isidorov et al., 1993).

Although environmental chloromethane is overwhelmingly biogenic in origin, what are the relative quantities of the other natural organochlorines and organohalogens vis-à-vis the approximately 20 million tons per year of 150 industrial organohalogens currently in use worldwide? Only a few studies have been carried out to address this question. A detailed study of the marine acorn worm, **Ptychodera flava** has revealed that a known population of these animals (64 million) excrete in their faecal matter 95 pounds of organohalogens daily in a one square kilometre habitat in Okinawa. Extrapolation of this to the worldwide population of marine worms leads to an enormous quantity of natural organohalogens from this one animal species (Higa and Sakemi, 1983). A study of natural chlorinated phenols, which are also produced by wood-rot fungi, has revealed that the concentrations of these natural compounds (up to 75 ppm) greatly exceeds the level of 10 ppm set by the Dutch government as being hazardous for man-made chlorophenols in soil! Likewise, the authors who demonstrated the natural production of perchloroethylene (PERC) and trichloroethylene (TCE) by marine algae conclude that 'the measured rates suggest that emission of PER and TCE for the oceans to the atmosphere may be of such a magnitude that it cannot be neglected in the global atmospheric chlorine budget'. Obviously, many more studies of a similar nature are needed in order to estimate the global quantities of natural organochlorines and other organohalogens.

Chlorinated phenols are found in ocean waters and in soil extracts. The major source of most of these compounds, such as 2,4,6-trichlorophenol, is biogenic and from natural halogenation processes such as reactions between humic acid and soil microbes (chloride and chloroperoxidase). Humic acids, which are degraded plant material, are not only present in soil, but also in rivers, lakes, and the oceans. Numerous studies have demonstrated the ease with which humic acid compounds, which are phenolic, react with chlorine to produce chloroform and other simple organochlorines via the natural breakdown of chlorophenols. Chloride is ubiquitous in soil, plants (up to 10 000 ppm), rivers, lakes, and, of course, oceans. These ingredients, in combination with natural chloroperoxidase and other enzymes, provide a plausible rationale for the natural formation of chlorinated phenols. Several studies have demonstrated that the natural production of chlorinated phenols outweighs their anthropogenic sources. The total pool of chlorinated compounds in peat bogs in Sweden is several hundred thousand tons, in areas of Sweden where these chemicals can only be of natural origin. By comparison, the largest industrial emissions are from paper pulp industries and are 10 000 tons per year.

Chlorinated humic acids have been found in ground water samples that date back 5200 years. The presence of these chlorinated compounds in pristine forests and lakes is a consequence of the natural chlorination of organic matter, probably as part of the normal mechanism of plant decay, and 70–75 per cent of the organochlorine substances in the River Rhine can only be explained by these natural chlorination processes. With the discovery of chlorinated humic acids in pristine areas far from industrial sources, one can estimate that, if the 0.24 per cent halogen content of one sample of soil humic acid is representative of the estimated 2.5 trillion tons of global humic acid, then our terrestrial and aquatic humic acid repository alone contains 6 billion tons of natural halogenated material. In any event, it seems likely that the natural biochlorination of humic material will prove to be a major source of chlorinated compounds in the environment. For example, the chloroform that is frequently found in groundwaters with no obvious anthropogenic source is probably due to the degradation of natural chlorinated humic acids.

The name 'dioxin' refers to a family of about 75 chlorine-containing chemicals, among them 2,3,7,8-tetrachlorodibenzo-p-dioxin, which is generally considered the most toxic of the group. Dioxin was first encountered as an impurity in some batches of the warfare defoliant Agent Orange. It is a by-product of some industrial processes that use chlorine, such as the bleaching of paper pulp, and is also produced during any combustion process, such as waste incineration, running motor-vehicle engines, steel-making and smelting, tobacco smoke, residential wood burning, and forest fires. Despite claims to the contrary over the past 20 years, dioxin is not the 'doomsday chemical of the twentieth century', nor is it the 'deadliest substance ever created by chemists'. Physicians and epidemiologists have been observing the health of the thousands of individual industrial workers, civilians, Vietnam veterans – who were exposed to dioxin at various levels during the past 40 years. None of these studies, described in more detail below, has been able to attribute unequivocally any human cancers or deaths to dioxin exposure. The only documented adverse health effect is the skin disease, chloracne. Although it is often persistent and disfiguring, chloracne is not life-threatening and is often reversible when exposure ceases. While dioxin tissue levels among Vietnam veterans in general are not significantly different (11.7 parts per trillion

[ppt]) from the levels of non-Vietnam veterans (soldiers who had never been to Vietnam, 10.9 ppt) or a civilian control group (12.4 ppt), certain groups of Vietnam veterans were exposed to higher levels of dioxin. Follow-up studies of these individuals show no association between dioxin tissue levels and cancer or other health effects. A two-part, 20-year mortality and health-effects evaluation of 995 Air Force Ranch Hands, the personnel who handled and sprayed Agent Orange, revealed that some had high tissue concentrations of dioxin (>300 ppt) 15 years after exposure. Among this group, there was no chloracne observed, no increase in nine immune-system tests and no increase in either melanoma or systemic cancer (cancers of the lung, colon, testicle, bladder, kidney, prostate; Hodgkin's disease; soft tissue sarcoma or non-Hodgkin's lymphoma). The authors of this 1990 study concluded that 'there is insufficient scientific evidence to implicate a causal relationship between herbicide exposure and adverse health in the Ranch Hand Group' (Wolfe et al., 1990; Michalek et al, 1990). Studies of more than 800 dioxin-exposed workers in nine industrial-plant accidents in the United States, England, Germany, France, Czechoslovakia and The Netherlands fail to indicate serious long-term health effects in these men, some of whom have dioxin concentrations exceeding 1000 ppt 30 years after their initial exposure. Some 465 cases of chloracne were observed in these workers. A study of 2200 Dow Chemical workers who were potentially exposed to dioxin revealed that they had a slightly lower mortality than a control group and that they have had no total cancer increase. A study of 370 wives of dioxin-exposed men showed no excess miscarriages and no excess foetal deaths or birth defects in their children. The Institute of Occupational Health at the University of Milan has published detailed evaluations of the human health effects of the July 1976 dioxin accident involving 37 000 people in Seveso, Italy (Bertazzi et al., 1989). Some of the exposed children in 'Zone A', the area of heaviest exposure, had dioxin tissue levels as high as 56 000 ppt immediately following the accident; but the only adverse health effect to date has been chloracne. Of the 193 cases of chloracne, 170 were in children under the age of 15; and the skin lesions in all but one of these cases had disappeared by 1985. Although it is essential to continue to monitor the health of the people in Seveso, the Institute report concluded that there were 'no increased birth defects due to dioxin exposure', since the children born during the period from 1977 to

1982 failed to demonstrate an increased risk of birth defects. The Seveso cancer mortality findings after 10 years do not allow firm conclusions. On the one hand, mortality from cancer of the liver, one of the organs targeted by dioxin, was no different from that of unexposed people; and breast cancer mortality tended to be below expectations. On the other hand, 'increases in biliary cancer, brain cancer, and lymphatic and haemopoietic cancer did not appear to result from chance. However, no definite patterns related to exposure classification were apparent'. In the words of a leading dioxin analyst, Dr. Michael Gough:

> No human illness, other than the skin disease chloracne, which has occurred only in highly exposed people, has been convincingly associated with dioxin. In short, epidemiological studies in which dioxin exposures are known to have been high, either because of the appearance of chloracne or from measurements of dioxin in exposed people, have failed to reveal any consistent excess of cancer. In those studies that have reported associations between exposure and disease, no chloracne was reported, and there are no measurements of higher-than-background levels of dioxin in the people who are classified as exposed (Gough, 1991).

Life-threatening health effects in humans have not been linked definitively to dioxin, despite the early fears to the contrary. Over 40 000 scientific papers have provided enormous information about this greatly misunderstood chemical, and the scientific and medical communities will continue to monitor the health of those people who have been exposed to large amounts of dioxins.

Some scientists believe that a large background source of dioxins in the environment is from forest fires, of which 200 000 occur annually, mostly caused by lightning. One study of the smoke from several Canadian forest fires and numerous laboratory studies of wood burning have confirmed that dioxins are produced under these conditions (Sheffield, 1985). Perhaps more surprising is that natural peroxidase enzymes convert chlorophenols into dioxins. These laboratory transformations lead to the formation of dioxins in the relatively high levels of 10–100 ppm! It has been demonstrated that dioxins are produced in fresh, uncontaminated garden compost piles and in municipal sewage sludge, presumably by the action of natural enzymes on organic material in the presence of chloride (Öberg et al., 1993).

Organochlorines, including dioxins, are found in ancient soil, water, peat, coal, and other organic samples, dating back 100, 1000, 4000, 35 000, 15 million, and even 300 million years ago! For example, halogenated fulvic acids have been isolated from groundwater samples that date back 1300, 4600, and 5200 years. Organohalogen material was identified in sediments dating back to the thirteenth century. Microfossils in Precambrian rocks, which are a billion years old, are identical to the blue-green alga **Nostoc**, and other microfossils are morphologically indistinguishable from **Oscillatoria**, two present-day species rich in organohalogen compounds. The analysis of pre-industrial glacial ice from Antarctica and northern Sweden has revealed trichloroacetic acid. Moreover, the latter compound, as well as chloroform, is found to be ubiquitous in soil, results confirmed by 37Cl-labelling experiments. The analysis of 35 000-year old organic matter (14C-dating) shows the presence of organochlorine, organobromine, and organoiodine compounds. Organochlorines have also been found in 1000-year old peat and 4000-year-old marine clay. Even more remarkable are the discoveries that organochlorines (200–300 ppm) have been detected in several-thousand-year-old peat from the Holocene period, in two lignite samples (107 and 166 ppm) that date from the Tertiary era, 15 million years ago, and in a 300-million-year-old bituminous coal sample (74 ppm) from the Upper Carboniferous period. Whether these natural organochlorines in sediments result from the deposition of biogenic material from plants already containing organochlorines or arise by the formation of organochlorines within the sediments remains to be established, although most of the evidence to date would support the former theory. These studies of sediments – ancient and modern – should provide information about the global quantities of natural organochlorines. Very recently, organochlorines have also been discovered in ancient meteorites (Nleusi et al., 1996). Thus, all indications are that organochlorine compounds have been with us since the formation of organic matter aeons ago and certainly organochlorine compounds predate the arrival of man on earth.

Numerous natural organochlorine compounds have potent antibacterial, anti-cancer, and other important medicinal properties. For example, the newly discovered anti-cancer Indian Ocean sponge metabolite, spongistatin, contains chlorine. An Oregon forest moss produces the chlorine-containing ansamitocin, which has potent anti-cancer activity against solid tumours, as does rebeccamycin, which contains two atoms of chlorine, discovered in a Panamanian soil microbe. Ambigol, a natural chlorinated PCB (polychlorinated biphenyl) from a terrestrial blue-green alga, is active against HIV reverse transcriptase which is an enzyme crucial for the AIDS virus. Recently, chloropeptin, which contains six atoms of chlorine, was discovered in a **Streptomyces** microbe. This remarkable chemical inhibits HIV replication in human blood cells. The well-known antifungal agent griseofulvin is another natural organochlorine compound, and the venerable microbial tetracycline antibiotics include at least ten examples of organochlorines, such as aureomycin. Another species of blue-green algae has yielded cryptophycin, a chlorine-containing compound that has unprecedented activity against solid tumours and which is almost certain to yield a new and important anti-cancer drug. Removal of the chlorine atom from cryptophycin results in a ten-fold decrease in potency. A soft coral from the Pacific Ocean produces a series of organochlorine prostaglandins, one of which is currently in the cancer clinic in Japan.

Eighty-five per cent of all synthetic pharmaceuticals and vitamins are made directly or indirectly through chlorine chemistry. Of the 400 new drugs approved for use in humans since 1984, more than 60 contain chlorine and many others contain one or more of the other halogens (bromine, fluorine, iodine). In 1993 three new organobromine drugs were introduced into medical practice. Of the top 100 United States prescription pharmaceuticals, 15 contain chlorine as an essential component of the molecule, and 13 others contain fluorine or iodine. Two of the ten most prescribed pharmaceuticals, ceclor and xanax, contain chlorine. Indeed, ceclor and the related organochlorine lorabid are the best drugs to treat middle-ear infections in children. Toremafene is used to treat breast cancer, and the naturally occurring antibiotic vancomycin is used to fight hospital staphylococcal infections that are resistant to penicillin. All of these drugs contain chlorine, as does the former pesticide DDD ('mitotane'), a DDT derivative used to treat inoperable adrenal cancer, cis-platin, the miracle testicular cancer drug, and the anaesthetic halothane. Chloroquine is an important synthetic antimalarial drug. Other chlorine-containing pharmaceuticals include claritin, ultravate, elocon, mepron, almide, wellbutrin, femstat, lopidine, selepam, melex, bonefos, aclovate, fareston, proendotel, propulsid, halfan, and sporanox. These drugs are used to treat millions of people annually for infections, fungal diseases, glaucoma, arthritis, inflammation, psoriasis, allergies, depression, osteoporosis, ulcers, malaria, coronary disease, and cancer. Numerous inhalation anaesthetics contain halogen, and two relatively new such organofluorine compounds are desfiurane and sevoflurane. Losartan is the new DuPont-Merck drug for hypertension, and CI-1002, the Parke-Davis agent for Alzheimer's Disease, and NE-10064, the new Procter and Gamble anti-arrhythmic drug are currently clinical candidates. Each of these is an organo-chlorine compound. It cannot be emphasised enough that the chlorine atoms in these molecules play an essential role in the function of these drugs. This is another way that chlorine saves lives.

Chlorine is used to purify drinking water and to disinfect swimming pools, both of which would otherwise be contaminated with faecal and other micro-organisms that cause diseases such as cholera, typhoid fever, dysentery, enteritis, giardiasis, hepatitis, and **E. coli** infections. Some 98 per cent of our public water systems are purified by chlorine or chlorine-based products. Chlorination is the water treatment of choice in preventing inestimable deaths every year. In the words of the director of the International Life Sciences Institute's Risk Science Institute, 'chlorination and disinfection of the water supplies are the public health success story of the century'. An estimated 80 per cent of all diseases and more than one-third of deaths in developing countries are caused by drinking or cooking with contaminated water. The World Health Organization estimates that worldwide 25 000 children die every day from waterborne diseases resulting from a lack of water disinfection. In Peru, the termination of water chlorination as a tragic experiment in 1991, because of a concern about chlorination by-products, resulted in a massive and unnecessary epidemic – causing more than one million cases of cholera and more than 20 000 deaths to date – that has since spread to 16 other Latin American countries (see Ghersi and Ñaupari Chapter 2).

Chlorine is used to bleach pulp in the manufacture of paper, recycled paper, and other paper products. The toxicity of chlorinated wastes from paper-pulp mills has been reduced significantly in recent years. A well-designed and well-maintained paper plant with secondary treatment removes up to 90 per cent of the effluent toxicity. Furthermore, many mills have switched to chlorine dioxide, a less reactive form of chlorine that produces fewer toxic by-products. The amount of chlorine used for pulp and paper bleaching in the United States fell from 1.4 million tons in 1990 to 920 000 tons in 1995, and the amount of chlorine dioxide used by the industry will increase from 670 000 to 975 000 tons over the same period. The dioxins that are produced and discharged to rivers and streams in small quantities when chlorine is used to bleach paper pulp – less than one pound annually from the whole USA paper industry – are exactly the same chemicals that are produced naturally in fires or by enzymatic action in living organisms as discussed earlier.

Many other sectors of society make use of organochlorine chemicals. Chloromethane is used in the production of silicones and as a solvent in butyl rubber polymerisation. Benzyl chloride is a raw material in the pharmaceutical industry and is used to synthesise phenobarbital, benzedrine, demerol, and other drugs such as those cited above. Vinyl chloride is polymerised to form polyvinyl chloride (PVC), a plastic polymer of incredible versatility and safety. This polymer is a building block for much of our manufacturing industry and an invaluable component of building materials, consumer goods, medical equipment, and many other products. For example, more than 500 million pounds of PVC are used annually for wire, cable, and other electrical applications. The presence of chlorine in PVC makes this material inherently flame retardant and, thus, ideal for construction and furnishing applications. Other uses of PVC include shoes, luggage, raincoats, paper and fabric coatings, computers and keyboards, food packaging, blood storage, garden hoses, credit cards, floor and wall coverings, recreational equipment, and more. Polytetrafluoroethylene – 'Teflon' – is made by the polymerisation of tetrafluoroethylene and because of its chemical stability has an amazingly diverse set of applications in our society.

Organohalogens, particularly organochlorine compounds, are used in crop protection. These chemicals have allowed people to enjoy safe, high-quality, insect-free food at much lower prices than would otherwise be possible. Some 96 per cent of all crop protectant pesticides and herbicides are chlorine-based. Ninety per cent of grain farms utilise these chemicals in food production. Examples that contain chlorine or bromine include the herbicides 2,4-dichlorophenoxyacetic acid (2,4-D) and picloram, and the soil fumigants bromomethane, 1,2-dibromoethane, and dibromochloropropane. Some other organohalogen pesticides are dursban, which is used in restaurants to kill cockroaches, daconil, a fungicide, bromoxynil, a rapidly biodegradable herbicide, and dyrene, a fungicide. Polychlorinated biphenyls (PCBs) were introduced in 1929 as insulators in the electric power industry in capacitors and transformers, as lubricants and coolants in vacuum pumps, as paint additives, in food packaging, and other applications. The manufacture of PCBs was discontinued in 1977 when it became clear that these materials were persistent in the environment and bioaccumulative with adverse effects on wildlife and humans.

As with all chemicals, 'the dose makes the poison', and this is true for the organochlorines. Even dioxin is not particularly toxic in some animals (dog, hamster, frog) but is highly toxic to others (guinea pig, horse). The insecticide DDT is highly effective in killing disease-ridden mosquitoes, ticks, and fleas, but is virtually non-toxic in mammals and has probably never killed a human being (see Kemm Chapter 1). In contrast, the chlorinated pesticides dieldrin, aldrin, endrin are highly toxic to mammals, and function as nerve poisons. None of these pesticides is in widespread use any longer.

The number of reported natural organochlorines from living organisms continues to increase, and these discoveries parallel scientific advances in collection, isolation, bioassay, and spectroscopic techniques. Given the plethora of known chlorinated marine natural products, it is surprising to realise that only a relatively small number of marine organisms have been investigated for their chemical content. Thus, whereas some 12 000 natural products of all types have been isolated from terrestrial plants, only 500 had been isolated from marine plants up to 1987. The fact that there are 500 000 species of marine animals, plants, and bacteria means that thousands of new organochlorine compounds are awaiting discovery, since the oceans are such a rich source of chloride (salt). For example, of the 4000 species of bryozoans (moss animals), fewer than a dozen have been examined for chemical metabolites. Similarly, the 3000 known species of opisthobranchs, 5000 sponges, 4000 segmented worms, and 80 000 molluscs each represent a fantastically rich and virtually untapped source of new organo-chlorine compounds. Moreover, exploration of marine bacteria and fungi is only just beginning, and this is a very promising area for future research.

Organochlorines cannot be banned – anymore than gravity or

photosynthesis can be banned – nor should the entire group of anthropogenic chlorinated chemicals be eliminated from our society because of the toxicity of a few. Obviously, we must monitor our output of all toxic chemicals, chlorinated or not. We must regulate organochlorines intelligently and with proper perspective, since nature – irrespective of our actions – will inexorably continue to churn out its own complement of organochlorines for its own purpose. Chlorine is as natural a component of our ecosystem as carbon, hydrogen, and oxygen.

General References

Gribble, G. W. (1994). 'The Natural Production of Chlorinated Compounds', *Environ. Sci. Tech.*, **28**, 310A.

Gribble, G. W. (1994). 'Natural Organohalogens-Many More Than You Think!', *I. Chem. Ed.*, **71**, 907.

Gribble, G. W. (1995). 'The Diversity of Natural Organochlorines in Living Organisms', *The Natural Chemistry of Chlorine in the Environment*, 5, First edition, Euro Chlor.

Gribble, G.W. (1995).'Chlorine and Health', American Council on Science and Health, Washington, D.C.

Gribble, G. W. (1996). 'Naturally Occurring Organohalogen Compounds – A Comprehensive Survey', *Progress in the Chemistry of Organic Natural Products*, **68**, 1.

Gribble, G. W. (1996). 'The Future of Chlorine', in Earth Day '96 – A Guide to Saving the Planet, http: I /www.heartland.org/earthday96/contents.htm

Gribble, G. W. (1996). 'The Diversity of Natural Organochlorines in Living Organisms', *Pure and Applied Chemistry*, **68**, 1699.

Gribble, G.W. (1998). 'Naturally Occurring Organohalogen Compounds', *Acct. Chem. Res.* **31**, 141.

Gribble, G.W. (1998). 'Chlorinated Compounds in the Biosphere, Natural Production', *Encyclopedia of Environmental Analysis and Remediation*, Meyers, R.A., Ed., John Wiley & Sons, New York, pp. 972–1035.

Specific References

Abrahamsson, K. et al. (1995). 'Marine Algae – A Source of Trichloroethylene and Perchloroethylene', *Limnol. Oceanog.*, *40*, 1321.

Bertazzi, P.A. et al. (1989). 'Ten-Year Mortality Study of the Population Involved in the Seveso Incident in 1976', *Am. J. Epidemiol.*, **129**, 1187.

Fowden, L. (1968). 'The Occurrence and Metabolism of Carbon-Halogen Compounds', *Proc. Roy. Soc. B*, **171**, 5.

Gough, M. (1991). 'Human Health Effects: What the Data Indicate', *Sci. Total Environ.*, **104**, 129.

Hazen, S.L. and Heinecke, J.W. (1997). '3-Chlorotyrosine, a Specific Marker of Myeloperoxidase-Catalyzed Oxidation, Is Markedly Elevated in Low Density Lipoprotein Isolated from Human Atherosclerotic Intima', *J. Clin. Invest.*, **99**, 2075.

Higa, T. and Sakemi, S. (1983). 'Environmental Studies on Natural Halogen Compounds, I: Estimation of Biomass of the Acorn Worm Ptychodera flava Eschscholtz (Hemichordata: Enteropneusta) and Excretion Rate of Metabolites at Kattore Bay, Kohama Island, Okinawa', *J. Chem. Ecol.*, **9**, 495.

Isidorov, V.A. (1990). *Organic Chemistry of the Earth's Atmosphere*, Berlin, Heidelberg; Springer-Verlag, Chapter 3.

Isidorov, V.A., Prilepsky, E.B., and Povarov, V.G. (1993). 'Photochemically and Optically Active Compounds of Minerals and Gas Emissions of Mining Plants', *J. Ecol. Chem.*, **N2–3**, 201.

Michalek, J.E., Wolfe, W.H., and Miner, J.C. (1990). 'Health Status of Air Force Veterans Occupationally Exposed to Herbicides in Vietnam, II: Mortality', *J. Am. Med. Assoc.*, **264**, 1832.

Sheffield, A. (1985). 'Sources and Releases of PCDDs and PCDFs to the Canadian Environment', *Chemosphere*, **14**, 811.

Nkusi, G. et al. (1996). 'Occurrence and Partial Chemical Characterization of Halogenated Organic Compounds in Carbonaceous Chondrites', *Abstracts of the V.M. Goldschmidt Conference*, Heidelberg, Germany; Vol. **1**, p. 435.

Öberg, L.G. et al. (1993). 'De Novo Formation of PCDD/Fs in

Compost and Sewage Sludge — A Status Report', *Organohalogen Cmpds.,* **11**, 297.
Symonds, R.B., Rose, W.I., and Reed, M.H. (1988). 'Contribution of Cl^- and F^--Bearing Gases to the Atmosphere by Volcanoes', *Nature,* **334**, 415.
Winterton, N. (1997). 'Are Organochlorine Compounds Created in the Human Body?', *Mut. Res.,* **373**, 293.
Wolfe, W.H. et al. (1990). 'Health Status of Air Force Veterans Occupationally Exposed to Herbicides in Vietnam, I: Physical Health', *J. Am. Med. Assoc.,* **264**, 1824.

9 Nature's hormone factory*

Jonathan Tolman

Summary

For millions of years plants have been producing chemicals, gradually perfecting a range of chemicals, some benign, some deadly. As the ability to detect, isolate, measure and test chemicals found in nature has progressed, a startling fact has emerged: hundreds of plants appear to produce endocrine disrupters – which affect hormonal balance in animals.

Many of the plants that produce phytoestrogens and other endocrine disrupters are edible. In laboratory tests, more than 43 plants and foods found in the human diet have been shown to be oestrogenically-active. Many phytoestrogen-containing plants are common elements of our diet. Such grains as corn and wheat form a significant part of the human diet. Many legumes have also shown a surprising capacity for phytoestrogen production.

Although much more work has been done on phytoestrogens, some work has been done on plant chemicals which are known to affect the production of sperm. Only a handful have been discovered in the human diet. Of these, the most common is cottonseed oil. Although cottonseed oil is rarely sold as a vegetable oil, it is commonly used in manufactured snack foods.

A great deal of attention has recently been given to the fact that synthetic chemicals have exhibited oestrogenic effects in laboratory studies. The chemicals most prominently cited are PCBs and DDT, both of which have been banned in most developed countries. Compared with phytoestrogens, the concern over synthetic oestrogens may be somewhat overstated. The oestrogenic effects from the phytoestrogens in our diet are an estimated 40 million times greater than those from synthetic chemicals. To date, however, there is no concrete evidence that either pose a risk to human health.

* A version of this chapter was published by the Competitive Enterprise Institute in 1996

Endocrine disrupters in the natural environment

The silent spring

In the 1940s the sheep ranchers of Australia began to notice a peculiar and frightening trend. At first there was a rash of stillborn lambs. Then the ewes became sterile. Each spring there were fewer and fewer lambs. For the ranchers it was literally a silent spring. By the mid-forties the sheep ranching industry in Australia was in a state of crisis and faced certain financial ruin unless the cause of the mysterious infertility in the sheep could be found. What could be causing this disastrous sterility. Genetic mutations? Radiation? Poisonous chemicals?

The Australian Department of Agriculture was called in: a cadre of veterinarians and scientists investigated all possible sources of the sterility. By 1946 they had discovered the source of the sheep's sterility – clover (Harborne 1991).

The ranchers were unaware that the innocuous-looking clover (*Trifolium subterraneum*) that they had recently begun planting in their fields to feed their sheep, had been producing large quantities of oestrogen-mimicking compounds. It took another decade before scientists finally pinned down the exact chemicals which were causing the sterility – genistein and formononetin (Bradbury and White 1954). A structural comparison for genistein and formononetin shows why they were causing the sterility. They are surprisingly similar in structure to oestrogen and synthetic oestrogen, diethylstilboestrol, DES. These two compounds mimic the steroidal nucleus of the natural female hormone, oestrogen. Although they are in fact rather weak oestrogens, the plants make up for that fact by producing them in comparatively huge quantities: 5 per cent of dry weight in the clover fodder (Harborne 1991).

The oestrogen-mimicking nature of chemicals found in plants isn't restricted to the clover *T. subterraneum*. A survey of clover found that 18 different species of the plant produced oestrogen-mimicking substances, or 'phytoestrogens', in quantities as high as those found in *T. subterraneum*.

Moreover, sheep are not the only animals on which the reproductive effects of phytoestrogens have been observed. Oestrogenic effects also have been observed in quail which feed on pastures rich in leguminous species (Leopold et al. 1976). In years of good rainfall, these legumes that are eaten grow luxuriantly and are relatively low in phytoestrogens. However, in drought years the levels of phytoestrogens are increased with respect to the weight of the leaves. Consequently, egg laying by female quails is curtailed. There appears to be a self-regulating mechanism whereby the increase in quail population is kept low when food availability is limited. In other words, in order to enhance their own survival, the plants reduce the number of quail in the next generation.

As the presence of phytoestrogens in clover and legumes became known, scientists began to wonder if other plants were also producing hormone mimickers which could disrupt the reproduction of animals, and even more importantly, whether any of these plants were in the human diet.

An answer to this question came from another area half way around the world, from Tibet, where scientists were struggling with a loss of fertility in sheep. A clue emerged from the fact that in the history of Tibet the population has been extremely stable, often for as long as 200 years. During those times the Tibetan diet largely consisted of barley and peas. Could the peas or barley be affecting the fertility of the Tibetans?

When scientists fed mice a diet consisting of 20 per cent peas, litter sizes dropped by 50 per cent. When the mice were fed diets consisting of 30 per cent peas, the mice failed to produce any young at all (Rensberger 1995).

The original chemical industry

There is now a large and growing body of research on plant chemicals (see Appendix A). Scientists are continually developing methods of detection, isolation, measurement, and testing of the prodigious number of chemicals that plants produce. Analysis of their effects has revealed a range of objectives. Some chemicals are used to directly deter animals from eating the leaves or seeds of the plants. Strychnine for example, is a very effective poison, found in the seed of **Strychnos toxifera**. Hundreds of plants produce an astounding array of powerful poisons, from exotic tropicals such as curare to the common flower foxglove, all deadly if eaten.

Producing deadly toxins is a rather straightforward method for a plant to avoid becoming lunch. Some plants, however, have chosen more subtle methods of chemical combat, not death but disruption – reproductive disruption.

Over the last few decades, scientists have been analysing plants and the chemicals they produce. Most of this research is aimed at finding new and useful drugs. One consequence of all these studies has been that a large number of plants have been studied for their reproductive effects. A survey of literature compiled by the University of Illinois at Chicago's Department of Medical Chemistry, showed that 149 chemicals have been isolated from plants and have demonstrated oestrogenic effects in laboratory studies. A total of 173 plants have been shown to have active oestrogenic effects (See Appendix A).

Although much research has focused on oestrogenic effects, anti-spermatogenic effects have also been studied. 55 compounds extracted from plants have been shown to inhibit the production of sperm, and 31 plant species have been tested and shown to be anti-spermatogenic (see Appendix B).

Lab testing of phytoestrogens

The reproductive effect of low levels of genistein has been tested in laboratory mice since the 1950s. Mice fed artificially elevated levels of genistein in the diet suffered decreased reproduction compared with control populations. In the control group, 82 per cent of the females produced litters. A second group was fed a diet which consisted of 0.2 per cent genistein. Of this second group, only 59 per cent of the females produced litters. A third group was fed commercial soya bean meal which contained 0.1 per cent genistein, of which 77 per cent of the females gave birth (Carter et al. 1958).

Other early studies have also been conducted on the oestrogenic effects of soya bean oil. In these tests, the oestrogenic effect was measured in terms of increased uterine weight in mice. One such study compared a host of different processed oils, both edible and non-edible. The oils were mixed into the feed to constitute 10 per cent of the diet. Of the two soya bean oil samples used, one increased the mean uterine weight by 77 per cent and the other by 86 per cent. Only two other oils produced more oestrogenic responses in the mice, coconut oil and rice bran oil (Booth et al. 1960).

Since these early studies, additional tests have continued to determine the reproductive effect of a number of phytoestrogens.

Many of the most recent studies have used new-born mice to assess how these substances affect the development of the reproductive tract. One such study, of another common phytoestrogen, coumestrol, found that: 'Neonatal coumestrol treatment is effective in causing a number of morphological alterations in the female reproductive tract' (Burroughs 1995).

Other studies have been conducted with other phytoestrogens in foetal and neonatal mice. One such study concluded, '[g]enistein influences oestrogen-dependent development by modifying both morphological and neuroendocrine endpoints.' In other words, phytoestrogens such as genistein and coumestrol, act as endocrine disrupters (Levy et al. 1995).

Phytoestrogens in food

Although hundreds of plants appear to produce endocrine disrupters, the concern for human exposure is limited principally to those plants in the human diet. Unfortunately, dozens of edible plants produce phytoestrogens and other endocrine disrupters. In laboratory tests more than 43 plants found in the human diet have been shown to be oestrogenically active (see Table 9.1).

Table 9.1: Common foods shown to have oestrogenic effects

Barley	Garlic	Pomegranate
Cabbage	Ginseng	Potato
Carob bean	Grapefruit juice	Rhubarb
Carrot	Green pea	Rice
Celery	Hops	Safflower oil
Chick pea	Kidney bean	Sage
Cinnamon	Lime	Sesame seed
Cloves	Liquorice	Soya bean
Coconut oil	Olive oil	Sugar beet
Coffee bean	Parsley	Sunflower oil
Common oat	Parsnips	Wheat
Corn	Peanut oil	Wild cherry
Cumin	Pineapple	Wild ginger
Fennel	Plum	Wild leek
Field bean		

Many phytoestrogen containing plants are common elements of our diet. Such grains as corn and wheat form a large part of our diet, for example. Many legumes have shown a surprising capacity for phytoestrogen production; it was not unexpected that phytoestrogens would be found in the legumes we eat.

But the soya bean has garnered particular attention. Soya beans are the third largest crop in the United States, Every year nearly 60 million acres of soya beans are planted in the United States alone, and nearly 2 billion bushels of soya beans are harvested.

Numerous studies have shown that soya beans contain significant levels of the phytoestrogens, genistein and daidzein. Dry soya beans on average contain 1107 milligrams of genistein per kilogram and 846 milligrams of daidzein per kilogram (Franke 1994). If one considers only the US crop of soya beans and only the amount found in the seeds, the total quantity of genistein produced each year is roughly 130 million pounds, and an additional 100 million pounds of daidzein.

Although some of the soya beans end up as animal feed, much of the genistein and daidzein is consumed by humans. When most people think of soya bean foods they generally think of tofu or soya sauce. But soya products are not restricted to a few occasionally eaten foods: they are ubiquitous in the modern western diet. Soya flour, soya protein and most importantly, soya bean oil are found in hundreds of products eaten every day.

The sprouting of the bean

At the turn of the century, soya beans were a virtually unknown crop in the United States. By 1924, when the US Department of Agriculture started publishing systematic annual statistics 448 000 acres of soya beans were harvested. Back then, the yield per acre was also low, a mere 11 bushels per acre. Since then, the acreage of soya beans harvested and the yield per acre have soared.

By the end of World War II, acreage had climbed to 10 million acres a year and the yield per acre had nearly doubled. As food processing modernised, more and more uses were being found for the soya bean, and demand continued to increase. In 1979, soya bean production peaked with 70 million harvested acres producing 2.3 billion bushels. Since then, production has continued to hover around 2 billion bushels a year.

The majority of soya beans grown today are crushed for their oil. Soya bean oil is by far the most common of the vegetable oils. Roughly 80 per cent of the vegetable oil consumed in the United States is soya bean oil. Annual per capita consumption of soya bean oil is an estimated 65 pounds per year (Chern 1991). For instance, most fried snack foods are manufactured with vegetable oil, predominantly soya bean oil. Virtually all infant formulas are a mixture of dairy and soya solids and proteins.

Antispermatogens found in food

Although much more work has been done on phytoestrogens, some work has been done on plant chemicals which are known to effect the production of sperm. Only a handful have been discovered in the human diet (see Table 9.2).

Of these, the most common is cottonseed oil, although it is rarely sold as a vegetable oil, it is used in many manufactured snack foods.

Since the 1970s most of the work on the antispermatogenic effects of cottonseed oil has been conducted in China, after Chinese authorities noted decreased fertility in some provinces in the late 1950s, and eventually linked it with the use of cottonseed oil in cooking (Liu 1957). Since then laboratory and human experiments have been conducted using cottonseed oil. In one study, rats fed 0.5 millilitres of cottonseed oil a day produced no viable sperm at the end of the 28-day test period (Huang 1980).

Table 9.2: Common foods from plants or containing chemicals shown to have antispermatogenic activity in laboratory experiments

Garlic	*Allium sativum* (Liliaceae)
Celery	*Apium graveolens* (Umbelliferea)
Cottonseed oil	*Gossypium hirsutum* (Malvaceae)
Sunflower seed oil	*Helianthus annuus* (Compositae)
Coffee	Caffeine (Alkaloid)
Tobacco	Nicotine (Alkaloid)
Grapefruit	Kaempferol (Flavonoid)
Chocolate	Theobromine (Alkaloid)
Tea	Theobromine (alkaloid)
Cola	Theobromine (alkaloid)

Xeno-oestrogens: cause for alarm?

A great deal of attention has recently been given to the fact that synthetic chemicals have exhibited oestrogenic effects in laboratory studies. The chemicals most prominently cited are PCBs and DDT, both of which have been banned in the United States. Compared with the phytoestrogens, the concern over synthetic oestrogens may be somewhat overstated.

Although they are ubiquitous in our diet, phytoestrogens are considered only weakly active. Most of the compounds which have been identified and tested have been found to have a relative potency when compared to synthetic oestrogen of 0.001 to 0.0001 (Safe 1995). In other words it takes between 1000 and 10 000 molecules of these phytoestrogens to create the same effect as one synthetic oestrogen molecule. The vast majority of phytoestrogens appear to occur in the family of chemicals known as flavonoids. Total flavonoid consumption in the human diet is estimated at approximately 1 gram per day (Harborne 1991). Consequently, the total daily oestrogenic effect of phytoestrogens could be estimated at roughly 100 micrograms of oestrogen equivalents (see Table 9.3).

Synthetic oestrogen-mimicking substances, or xeno-oestrogens are considered even less oestrogenically active than the phytoestrogens. Oestrogenically active pesticides such as DDT, Dieldrin, and Endosulfan have been assigned a relative potency of 0.000001 (Soto et al. 1994). In other words it takes one million xeno-oestrogen molecules to have the same effect as one synthetic oestrogen molecule. Since total intake of these xeno-oestrogens is significantly lower than the naturally occurring phytoestrogens, (around 2.5 micrograms per day) the oestrogen exposure from synthetic chemicals has been estimated at 0.0000025 micrograms of oestrogen equivalents per day (Safe 1995). In other words, the estimated oestrogenic effects from the phytoestrogens in our diet

Table 9.3: Natural vs synthetic oestrogens

Source of oestrogen	Oestrogen equivalent (μg/day)
Birth control pill	18 675
Postmenopausal therapy	3350
Phytoestrogens in food	102
Xeno-oestrogens	0.0000025

are 40 million times greater than those from synthetic chemicals, but it is questionable that either affect human health.

It is widely recognised that results obtained in vitro (in a test tube) are not necessarily reproduced in vivo (in the body). For example, oestradiol, taken by mouth has no activity, it simply passes through the body and is excreted; it is only in an 'active' form (such as ethynyloestradiol – a constituent of the birth control pill), that it can be absorbed by the body. However, a screening test on oestradiol in vitro would show a high degree of activity. This is why we need to be cautious of such tests.

Plant defence theory

A number of hypotheses have been developed to explain the functioning of plant defensive systems. Two of these hypotheses in particular bear examination. Because defences are costly to produce:

1. Less well-defended individuals have greater fitness than more highly-defended individuals, when enemies are absent.
2. Commitment to defence is decreased when enemies are absent and increased when attacked.

A sizeable body of literature has been accumulated which supports both these hypotheses (Rosenthal and Janzen 1979). For example, varieties of insect resistant soya bean produce a lower yield of seeds and accumulate nitrogen more slowly than insect susceptible varieties in the absence of herbivores including insects.

In the case of the second hypothesis, numerous studies have been conducted actually measuring the increased amounts of secondary metabolites. In the species *Senecio jacobaea* (Toyon), when half of the leaves were removed the plant responded within two days by increasing the amount of total leaf alkaloids and N-oxides in the remaining leaves, 40 to 47 per cent.

In another case, when beet plants were infested by beet flies, within 24 days the mortality of the beet flies increased between 29 and 100 per cent. The ability of beets to respond to infestation, and in some cases kill all of the infesting insects, shows the extreme effectiveness of some secondary plant metabolites and the plant's ability to defend itself (Rosenthal and Janzen 1979, and see Appendix A).

Risk v. risk analysis for crop protection

Current US government policy only rarely regulates naturally-occurring chemicals in the food supply. In only two cases has the US Food and Drug Administration (FDA) directly regulated naturally-occurring carcinogens. In the first instance, the FDA banned the chemical, safrole, which had been used as a natural flavouring for root beer, after it was determined that safrole was highly carcinogenic in high-dose rat studies.

The second case of federal regulation of a natural chemical is aflatoxin. The FDA has established a standard of 20 parts per billion of aflatoxin for a variety of foods susceptible to aflatoxin poisoning, most commonly peanuts (*Food Chemical News* 1991). In setting its aflatoxin standard, the FDA engaged in a type of risk-versus-risk analysis.

The FDA recognised that it would be virtually impossible to eliminate aflatoxin from the food supply. To do so would have required the massive use of fungicides which may present other equally hazardous long term health effects. The question which faced the agency was, what level of aflatoxin contamination should be tolerated to maximise human health? Whether or not the Agency came to the correct answer is difficult to determine. However, the fact that the Agency considered the health effects of both aflatoxin and prophylactic fungicides increases the chance of establishing a better health standard.

This type of risk-risk analysis could benefit other regulatory programmes affecting agriculture and the production of food. Current pesticide regulation, for example, does not take into account the potential of pesticides applied on crops to inhibit the formation of the plant's own secondary metabolites.

More importantly, as biotechnology and the ability to manipulate the genetic code of plants increases, the effects of secondary plant metabolites may increase geometrically.

A final note

More than fifty years after discovering the fertility problems in sheep in Australia, ranchers continue to struggle with clover disease. Despite decades of research and the attempted introduction of more benign types of clover, clover disease is still responsible for the loss of 1 million lambs every year.

Had clover disease struck herds of sheep in the middle ages or even in the seventeenth century the source of this infertility would likely have been blamed on a witch. And no doubt, if the infertility continued, the shepherds would likely have rounded up some socially unpopular woman with few friends to defend her, held a mock trial, called in religious authorities to absolve their collective conscience and then burned her at the stake.

Although science has done much to dispel myths and witches, there still exists in humans the urge to blame the socially unpopular for the ills of humanity. In the modern sense it is easy for us to blame a witch, chemical or otherwise. But the responsible, and far more laborious, task is to determine what actually is causing the problem, which is often made even more difficult when we discover, as the sheep ranchers in Australia have, that even decades of science can't always solve the problem.

References

Booth, A. N. Bickhoff, E. M. and Kohler, G. O. (1960) 'Oestrogen-like Activity in Vegetable Oils and Mill By-products', *Science*, June: 1807–8.

Burroughs, C. D. (1995). 'Long-Term Reproductive Tract Alterations in Female Mice Treated Neonatally with Coumestrol', *Proceedings Of the Society For Experimental Biology and Medicine*, **208**, 1:79 January.

Bradbury, R. B. and White, D. E. (1954). *Vitam. Horm.*, 12:207–233.

Carter, M. W., Matrone, G. and Smart, Jr W. W. G. (1958). 'Effect of Genistein on Reproduction of the Mouse', *Journal of Nutrition*, **55**, p. 639–645.

Chern, W. S. (1991). 'Health Risks and Oil Consumption in the American Diet', Ohio State University, United Press International, June 9.

Food Chemical News, (1991). 'Aflatoxin Controls Effective, GAO Tells Congress', May 27.

Franke, A. A. (1994). 'Quantitation of Phytoestrogens in Legumes by HPLC', *Journal of Agricultural and Food Chemistry* **42**, 1905–1013.

Harborne, J. B. (1979). 'Flavonoid Pigments', *Herbivores: Their Interactions with Secondary Plant Metabolites,* Edited by Gerald A. Rosenthal and May Berenbaum. Academy Press.

Huang, L. (1980). 'Chemical Studies of Gossypol', Proceedings of World Health Organisation – National Institutes of Health Meeting

of the Core Group of Advisors To The Chemical Synthesis Program NIH Bethesda, Nov. 17–18.
Leopold, A. S., Erwin, M., Oh, J. and Browning, B. (1976). *Science* **191**:98–100.
Levy, J. R. et al. (1995). 'The Effect of Prenatal Exposure to the Phytoestrogen Genistein on Sexual Differentiation in Rats', Proceedings of the Society for Experimental Biology and Medicine, 208: 1, January.
Liu, S. B. (1957). 'Suggestion of Feeding on Crude Cottonseed Oil For Contraception', *Shanghai Acta Medica* **6**:43.
Rensberger, B. (1995). 'Conception The Natural Way', *Washington Post*, July 25, Science A3.
Rosenthal, G. A. and Janzen, D. H. (1979). *Herbivores, Their Interaction with Secondary Plant Metabolites*, Edited by Gerald A. Rosenthal and May Berenbaum, Academy Press. p 14.
Safe, S. (1995). 'Environmental and Dietary Oestrogens and Human Health: Is There A Problem?' *Environmental Health Perspectives*, **103**, 4:349, April.
Soto, A., Kerrie, M., Chung, L. et al. (1994). 'Pesticides Endosulfan, Toxaphene and Dieldrin Have Oestrogenic Effects on Human Oestrogen Sensitive Cells', *Environmental Health Perspectives*, **102**, 4:381.

Appendix A

Biological activities for extracts of plants with estrogenic activity compiled from University of Illinois at Chicago natural products alert database

Plant scientific names are listed along with an abbreviated citation. Full citations available from the author upon request.

Abroma augusta (Sterculiaceae) leaf India	Riv Zootec Vet 1976: 247- (197
Abroma augusta (Sterculiaceae) root India	Riv Zootec Vet 1976: 247- (197
Abutilon indicum (Malvaceae) dried fruit India	Phytother Res 32: 61-66 (1989)
Achyranthes aspera (Amaranthaceae)	Curr Sci 54 22:1197-1199 (1985
Aegopodium podagraria (Umbelliferae)	Can J Anim Sci 60: 531-534 (19
Agave americana (Agavaceae)	Rev Espan Fisiol 16 1:7-(1960)
Allium ampeloprasum (Liliaceae)	Rev Espan Fisiol 16 1:7- (1960)
Allium sativum (Liliaceae) bulb	Nord Vet Med 25: 91-96 (1973)
Allium sativum (Liliaceae)	Aust J Agr Res 19: 1069-1076
Alnus glutinosa (Betuilaceae) leaf USA	Endocrinology 14: 389- (1930) English
Alnus glutinosa sex female (Betulaceae)	Endocrinology 14: 389- (1930) English
Alnus glutinosa sex male (Betulaceae)	Endocrinology 14: 389- (1930) English
Aloe indica (Liliaceae)	Indian J Pharmacol 12: 57-58
Althaea rosea (Malvaceae)	Biochem Z 180: 11- (1927)
Althaea rosea (Malvaceae)	Biochem Z 180:11- (1927)
Anagallis arvensis (Primulaceae)	Rawal Med J 51:21-26 (1976)
Anagallis arvensis ssp.arvensis (Primulaceae)	Bangladesh J Biol Agr Sci 4: 6
Ananas comosus (Bromeliaceae) fruit USA-FL	Science 121:42- (1955)
Androsace septentrionalis (Primulaceae)	Korean J Pharmacog 15 2: 85-90
Anthericum liliago (Liliaceae)	Rev Espan Fisiol 16 1: 7- (1960)
Arachis hypogaea (Leguminosae) seed oil USA	Science 131:1807- (1960)
Artocarpus integrifolia (Moraceae)	Indian J Physiol Allied Sci 20: 6- (1967)
Asclepias tuberosa (Asclepiadaceae) root USA-GA	J Amer Pharm Ass Sci Ed 39: 233- (1950)
Asclepias tuberosa (Ascilepiadaceae)	Nord Vet Med 25:91-96 (1973)
Asclepias tuberosa (Ascilepiadaceae)	Aust J Biol Sci 28: 279-290 (1
Asparagus racemosus (Liliaceae)	Int Conf Chem Biotechnol Biol A
Astragalus lentiginosus (Leguminosae)	Can J Comp Med 36: 360- (1972)
Astragalus miser var. serotinus (Leguminosae)	Mie Daigaku Nogakubu Gakujutsu
Avena sativa (Gramineae)	Nord Vet Med 25: 91-96 (1973)
Avena sativa (Gramineae) seed	Hoppe-Seyler's Z Phys Chem 218: 104- (1933)
Avena sativa (Gramineae) seed	Endocrinology 14: 389- (1930) English
Avena sativa (Gramineae) seedling	Hoppe-Seylers Z Phys

	Chem 218: 104- (1933)
Barbarea vulgaris (Cruciferae)	Can J Anim Sci 60: 531-534 (19
Beta vulgaris (Chenopodiaceae)	Planta Med 17: 71- (1969)
Beta vulgaris (Chenopodiaceae)	J Sci Ind Res-C 15: 202-204 (1
Brassica oleracea (Cruciferae)	Experientia 16:194- (1960) English
Brassica oleracea (Cruciferae) leaf	Comp Physiol Ecol 8 2:101-104
Breynia patens (Euphorbiaceae)	Indian J Pharmacol 12: 57-58
Butea monosperma (Leguminosae)	Patent-US-2,136,397 (1938)
Butea monosperma (Leguminosae)	Patent-Ger-651,857 (1937)
Butea monosperma (Leguminosae)	Abstr 10th Conference of Science
Butea monosperma (Leguminosae)	J Ethnopharmacol 25 3: 249-261
Butea monosperma (Leguminosae)	Patent-US-2,112,712 : 2pp- (19
Calendula officinalis (Compositae)	J Anim Sci 15: 25- (1956)
Calendula officinalis (Compositae)	Poznan Towarz Przyjaciol Naukw
Carthamus tinctorius (Compositae) seed oil USA	Science 131:1807- (1960)
Cassia fistula (Leguminosae) dried fruit India	Dokl Bolg Akad Nauk 33:1565-1
Ceratonia siliqua (Leguminosae) seed oil	Lek Sredtsva Daunego Vostoka 7
Chenopodium album (Chenopodiaceae)	Farmakol Prir Veschestv 1978:
Chenopodium album (Chenopodiaceae)	Can J Anim Sci 60: 531-534 (19
Cicer arietinum (Leguminosae) seed oil India	Mater Biol Vidov Roda Glycyrri-11
Cinnamomum zeylanicum (Lauraceae) bark	J Amer Vet Med Ass 134: 237-
Cinnamomum zeylanicum (Lauraceae) bark	Anais Assoc Brasil Quim 2: 57
Cocos nucifera (Palmae) seed oil	Science 131:1807- (1960)
Coffea arabica (Rubiaceae)	Proc Amer Vet Med Ass 19: 150
Combretodendron africanum (Lecythidaceae)	Fitoterapla 60 6: 547-548 (1989
Commiphora myrrha (Burseraceae)	Indian J Pharmacol 12: 57-58
Convallaria majalis (Liliaceae)	Vitamins And Hormones 1954: 207- (1954)
Costus speciosus (Zingiberaceae)	Indian J Med Res 60: 287- (1972)
Costus speciosus (Zingiberaceae)	J Anat Soc India 20:136- (1971)
Costus speciosus (Zingiberaceae)	Indian J Pharmacy 34: 116- (1972)
Crocus sativus (Iridaceae)	Yao Hsueh Hsueh Pao 112: 94-100 (1964)
Cuminum cyminum (Umbelliferae) dried seed	Fitoterapia 58 1: 9-22 (1987) E
Cynodon dactylon (Gramineae)	J Res Indian Med Yoga Homeopath
Cynodon dactylon (Gramineae)	Rhod Zambia Malawi J Agr Res 5:179- (1967)
Cyperus rotundus (Cyperaceae)	Arch Inst Farmacol Exp(Madrid)
Cytisus scoparius (Leguminosae)	Physiology of Reproduction 1963
Dactylis glomerata (Gramineae)	J Amer Vet Med Ass 118: 323- (1951)
Dactylis glomerata (Gramineae)	J Roy Egypt Med Ass 30: 124-12
Dactylis glomerata (Gramineae)	Food Cosmet Toxicol 18:

	425-42
Dactylis glomerata cv. Danish (Gramineae)	Izv Selskokhoz Nauk 15 12: 53
Datura quercifolia (Solanaceae)	Planta Med 1986 6: 552-B (1986)
Daucus carota (Umbelliferae) root France	C R Seances Soc Biol Ses Fil 157:1024- (1963)
Daucus carota (Umbelliferae)	Indian J Pharmacol 13: 60, (1
Daucus carota (Umbelliferae)	Planta Med 1986 6: 552-B (1986)
Daucus carota (Umbelliferae)	Comp Physiol Ecol 8 2:101-104
Dianthus superbus (Caryophyllaceae)	Int Z Klin Pharma Ther Toxikol 2: 366- (1969)
Digitalis lanata (Scrophulariaceae)	Indian J Pharmacol 13 1: 67-68
Digitalis lanata (Scrophulariaceae)	Nord Vet Med 32: 480-486 (1980
Duchesnea indica (Rosaceae)	IRCS Med Sci 11 : 522, (1983)
Elaeis guineensis (Palmae)	Hoppe-Seylers Z Phys Chem 218:104- (1933)
Eleutherococcus senticosus (Araliaceae)	Nutr Bromatol Toxicol 5 2:67-7
Eleutherococcus sessiliflorus (Araliaceae)	Ginseng and Other Drugs of Soviet Far E.
Eleutherococcus spinosus (Araliaceae)	Arab Sci Congr 5th, Baghdad 196
Embella ribes (Myrsinaceae)	J Med Ass Thailand 24 2: 83-94
Embelia ribes (Myrsinaceae)	Nippon Chikusan Gakkaiho 49: 2
Ensete superbum (Musaceae)	Planta Med 1986 6: 552-B (1986)
Eragrostis curvula (Gramineae)	Boll Soc Ital Biol Sper 14: 83
Ervatamia heyneana (Apocynaceae)	J Pharm Sci 62: 1199- (1973)
Ervatamia heyneana (Apocynaceae)	Planta Med 1986 6: 552-B (1986)
Eucalyptus globulus (Myrtaceae)	Biochem J 32: 641-645 (1938) English
Euphorbia hirta (Euphorbiaceae)	Kunitsa, Lk
Ferula involucrata (Umbelliferae)	Farmakol Prir Veschestv 1978:
Ferula jaeschkeana (Umbelliferae)	Farmakol Prir Veschestv 1978:
Ferula jaeschkeana (Umbelliferae)	J Pharm Pharmacol Suppl 41:18
Ferula jaeschkeana (Umbelliferae)	Planta Med 1986 6: 552-B (1986)
Ferula ugamica (Umbelliferae)	Farmakol Prir Veschestv 1978:
Ficus infectoria (Moraceae)	Indian J Physiol Allied Sci 20: 6- (1967)
Ficus religiosa (Moraceae)	Indian J Physiol Allied Sci 20: 6- (1967)
Foeniculum vulgare (Umbelliferae)	Biochem J 32: 641-645 (1938) English
Foeniculum vulgare (Umbelliferae)	Shoyakugaku Zasshi 39 1: 46-51
Genista rumelica (Leguminosae)	Indian Vet J 57: 762-765 (1980
Glyceria grandis (Gramineae)	Can J Anim Sci 60: 531-534 (19
Glycine max (Leguminosae)	Nutrition 42: 487- (1950)
Glycine max (Leguminosae)	J Nutr 42: 487- (1950)
Glycine max (Leguminosae)	J Endocrinol 24: 341-348 (1962
Glycine max (Leguminosae)	Philippine Agr 46: 673- (1963)
Glycine max (Leguminosae) seed oil USA	Science 131:1807- (1960)
Glycine max (Leguminosae)	Food Cosmet Toxicol 18: 425-42
Glycine max (Leguminosae) dried seed	Farmatsiya (Sofia) 32 2: 52-58
Glycine max (Leguminosae) dried seed	Comp Physiol Ecol 8 2:101-104

Glycyrrhiza glabra (Leguminosae)	Rast Resur 8: 490- (1972)
Glycyrrhiza glabra (Leguminosae)	Norsk Farm Selskap 12: 68-73
Glycyrrhiza glabra (Leguminosae)	Aktual Vopr Farm 1970:112- (1970) Russian
Glycyrrhiza glabra (Leguminosae)	Zentralbl Veterinaermed 10A: 155- (1963)
Glycyrrhiza glabra (Leguminosae)	Zentralbl Veterinaermed 11A: 70- (1964)
Glycyrrhiza glabra (Leguminosae)	Zentralbl Veterinaermed 11A: 773- (1964)
Glycyrrhiza glabra (Leguminosae)	J Endocrinol 31 :289- (1965)
Glycyrrhiza glabra (Leguminosae)	Patent-Ger-649,202 :- (1937)
Glycyrrhiza glabra (Leguminosae)	Pharmazie 25: 620- (1970)
Glycyrrhiza glabra (Leguminosae)	J Amer Pharm Ass Sci Ed 39:177- (1950)
Glycyrrhiza glabra (Leguminosae)	Rast Resur 7: 295- (1971)
Glycyrrhiza uralensis (Leguminosae)	Norsk Farm Selskap 12: 68-73
Haplophyllum latifolium (Rutaceae)	Farmakol Prir Veschestv 1978:
Haplophyllum perforatum (Rutaceae)	Planta Med 36: 369-374 (1979)
Haplophyllum perforatum (Rutaceae)	Farmakol Prir Veschestv 1978:
Helianthus annuus (Compositae) seed	Vitamins and Hormones 1954: 207- (1954)
Heracleum sosnowskyi (Umbelliferae)	Rast Resur 7: 295- (1971)
Hibiscus sabdariffa Cv. El-Fasher (Malvaceae)	Nat Acad Sci Lett 11 3: 81-84
Hordeum vulgare (Gramineae)	J Anim Sci 17: 787-791 (1958)
Humulus lupulus (Cannabaceae)	Experientia 16:194- (1960) English
Humulus lupulus (Cannabaceae)	Brauwissenschaft 14: 4- (1961)
Humulus lupulus (Cannabaceae)	Amer Perf Aromat 75 5: 61-62 (1960) English
Humulus lupulus (Cannabaceae)	Comp Physiol Ecol 8 2:101-104
Humulus lupulus (Cannabaceae)	Muench Med Wochenschr 95: 845- (1953)
Humulus lupulus (Cannabaceae)	Suddeut Apoti-1-Ztg 78: 645- (19
Humulus lupulus (Cannabaceae)	Nord Vet Med 25: 91-96 (1973)
Hyparrhenia filipendula (Gramineae)	Rhod Zambia Malawi J Agr Res 5: 179- (1967)
Hyptis suaveolens (Labiatae)	Theriogenology 19 4: 507-516 (1
Hyptis suaveolens (Labiatae)	Planta Med 1986 6: 552-B (1986)
Ilex cornuta (Aquifoliaceae)	Turrialba 24 2:147-156 (1984)
Impatiens parviflora (Balsaminaceae)	Biochem Z 180:1- (1927)
Impatiens parviflora (Balsaminaceae)	Biochem Z 180:1- (1927)
Impatiens parviflora (Balsaminaceae)	Vitamins And Hormones 1954: 207- (1954)
Inula grandis (Compositae)	Farmakol Prir Veschestv 1978:
Ixora finlaysoniana (Rubiaceae)	Contraception 44 5: 549-557 (19
Leonurus sibiricus (Labiatae)	J Endocrinol 12: 261-266 (1955) English
Leonurus sibiricus (Labiatae)	Pharmacol Rev 11 : 135-172 (195
Lepidium capitatum (Cruciferae)	Vet Med Nauki 14 8:69-75 (1977
Leptadenia reticulata (Asclepiadaceae)	Indian J Pharmacol 12: 57-58 (
Lespedeza bicolor (Leguminosae)	Crop Sci 5: 276- (1965) English

Leucaena glauca (Leguminosae)	Arch Int Pharmacodyn Ther 78:
Levisticum officinale (Umbelliferae)	J Sci Ind Res-C 14:128- (1955
Linum usitatissimum (Linaceae) seed oil	Science 131:1807- (1960)
Lithospermum ruderale (Boraginaceae)	Nippon Sochi Gakkaishi 17 3:20
Lithospermum ruderale (Boraginaceae)	Physiology of Reproduction, F.L
Lolium perenne (Gramineae)	J Roy Egypt Med Ass 30: 124-12
Lolium rigidum (Gramineae)	J Amer Pharm Ass Sci Ed 41: 32
Lotus corniculatus (Leguminosae)	Publ Amer Ass Adv Sci 53: 195- (1959)
Lotus corniculatus (Leguminosae)	Patent-Brit-437,051:- (1935)
Lupinus albus (Leguminosae)	Planta Med 12:155- (1964)
Lupinus angustifolius cv.Tifblue-78 (Leguminosae)	J Ethnopharmacol 29 1:13-23 (1
Lycium chinense (Solanaceae)	Bull Agr Congo Belge 47: 1345
Lycopodium clavatum (Lycopodiaceae)	Farmacognosia 18:179- (1958)
Malus sylvestris (Rosaceae)	Vitamins and Hormones 1954: 207- (1954)
Malvaviscus conzattii (Malvaceae)	Indian J Med Res 1988 4: 336-35
Mangifera indica (Anacardiaceae)	Indian J Physiol Allied Sci 20: 6- (1967)
Medicago sativa (Leguminosae)	Bull Haffkine Inst 3: 346-. (1
Medicago sativa (Leguminosae)	Ann Endocrinol 29: 699- (1968)
Medicago sativa (Leguminosae)	Proc XI Ann Conf Indian Pharmac
Medicago sativa (Leguminosae)	Vet Med Nauki 14 8:69-75 (1977
Medicago sativa (Leguminosae)	Indian J Exp Biol 3: 61-63 (19
Medicago sativa (Leguminosae)	J Anim Sci 18: 1000-1008 (1959
Medicago sativa (Leguminosae)	Publ Amer Ass Adv Sci 53: 195- (1959)
Medicago sativa (Leguminosae)	J Dairy Sci 39: 81- (1955)
Medicago sativa (Leguminosae)	Tohoku J Exp Med 106: 21 G- (19
Medicago sativa (Leguminosae)	Publ Amer Ass Adv Sci 53: 195- (1959)
Medicago sativa (Leguminosae)	Publ Amer Ass Adv Sci 53: 195- (1959)
Medicago sativa (Leguminosae)	Diss Abstr Int B 23: 2286- (1963) English
Medicago sativa (Leguminosae)	Rast Resur 7: 295- (1971)
Medicago sativa (Leguminosae)	Science 126: 969- (1957)
Medicago sativa (Leguminosae)	J Reprod Fertil 6:115- (1963)
Medicago sativa (Leguminosae)	Can J Anim Sci 60: 531-534 (19
Medicago sativa (Leguminosae)	Food Cosmet Toxicol 18: 425-42
Medicago sativa (Leguminosae)	Poznan Towarz Przyjaciol Nauk, W
Medicago sativa (Leguminosae)	Rast Resur 12: 515-525 (1976)
Medicago sativa (Leguminosae)	Indian J Pharm Sci 45 2: 97-99
Medicago sativa (Leguminosae)	Vet Med Nauki 13 6: 59-65 (1976
Medicago sativa (Leguminosae)	Phytother Res 1 4:154-157 (198
Medicago sativa (Leguminosae)	J Anim Sci 16: 850-853 (1957)
Medicago sativa (Leguminosae)	Biochem Z 291 : 263-265 (1937)
Medicago sativa (Leguminosae)	Agricultura 10 2: 353-358 (1962
Medicago sativa cv. Dunavka (Leguminosae)	Rrl Jammu Newsletter 6 2:6-7
Medicago sativa cv. Dupuit (Leguminosae)	Izv Selskohoz Nauk 15 12: 53
Medicago sativa ssp. sativa (Leguminosae)	Rev Cresterea Anim 29 2: 52-59
Melilotus alba (Leguminosae)	Publ Amer Ass Adv Sci 53:

	195- (1959)
Moringa pterygosperma (Moringaceae)	Indian J Exp Biol 25 7: 442
Moringa pterygosperma (Moringaceae)	Elsevier Science Publishers B.V
Morus alba (Moraceae)	Indian J Physiol Allied Sci 20: 6- (1967)
Musa species (Musaceae)	Kachiku Hanshokligaki.1 Zasshi 23
Musanga cecropioides (Moraceae)	An Farm Bioquim (Buenos Aires) 2
Nuphar luteum (Nymphaeaceae)	Vitamins and Hormones 1954: 207- (1954)
Nuphar luteum (Nymphaeaceae)	Biochem Z 180:1- (1927)
Nuphar species (Nymphaeaceae)	Endocrinologie 1: 411-418 (192
Olea europaea (Oleaceae) seed oil	Science 131:1807- (1960)
Onobrychis sibirica (Leguminosae)	Sbornik Ceskoslov Akad Zemedel
Oryza sativa (Gramineae)	J Sci Food Agr 14: 376- (1963)
Oryza sativa (Gramineae) seed	Maslo-Zhir Prom-St 37: 23-24
Oryza sativa (Gramineae) seed oil	Science 131:1807- (1960)
Oryza sativa (Gramineae) dried seed	Comp Physiol Ecol 8 2:101-104
Panax ginseng (Araliaceae) root	Acta Pharm Int 1: 243- (1950)
Peganum harmala (Zygophyllaceae)	Indian J Pharmacol 12: 57-58
Petroselinum crispum (Umbelliferae)	Vitamins and Hormones 1954: 207- (1954)
Phaseolus vulgaris (Leguminosae) seed	Planta Med 11 :450- (1963)
Phaseolus vulgaris (Leguminosae) dried seed	Kgl- Lantbruks-Hogskel Ann 26:
Phleum pratense (Gramineae)	Ostrovsky, D: Kitts, WD
Phleum pratense (Gramineae)	J Reprod Fertil 6:115- (1963)
Phleum pratense (Gramineae)	Indian J Pharmacol 12: 60a- (1
Phleum pratense cv. climax (Gramineae)	Izv Selskokhoz Nauk 15 12: 53
Phoenix dactylifera (Palmae) kernel	Z Physiol Chem 218:104- (1933) German
Phoenix dactylifera (Palmae)	J Egypt Med Ass 30: 124- (1947)
Phoenix dactylifera (Palmae)	Nature (London) 159: 409- (1947)
Phoenix dactylifera (Palmae)	Nord Vet Med 25: 91-96 (1973)
Phoenix dactylifera (Palmae)	Proc Soc Exp Biol Med 73: 311
Pimpinella anisum (Umbelliferae)	J Endocrinol 31 :289- (1965)
Pimpinella anisum (Umbelliferae)	Biochem J 32: 641-645 (1938) English
Pimpinella anisum (Umbelliferae)	Patent-Ger-649,202:- (1937)
Pinus ponderosa (Pinaceae)	Can J Anim Sci 41:1- (1961) English
Pinus ponderosa (Pinaceae)	Feedstuffs 51 :54-68 (1979) English
Piper longum (Piperaceae)	Vet Sci 13: 59.65 (1976) Russian
Pisum sativum (Leguminosae)	Experientia 16:194- (1960) English
Pisum sativum (Leguminosae) seed	Planta Med 11 : 450- (1963)
Pisum sativum (Leguminosae) seed	Comp Physiol Ecol 8 2:101-104
Polygonum ochoreum (Polygonaceae)	Can J Anim Sci 60: 531-534 (19
Prunus avium (Rosaceae) fruit pulp	Vitamins And Hormones 1954: 207- (1954)

Prunus domestica (Rosaceae) fruit pulp	Vitamins And Hormones 1954: 207- (1954)
Psoralea corylifolla (Leguminosae)	J Endocrinol 12: 261-266 (1955) English
Psoralea corylifolla (Leguminosae)	J Agr Sci 83: 505- (1974)
Psoralea corylifolla (Leguminosae)	Res Vet Sci 22: 216-221 (1977)
Pueraria mirifica (Leguminosae)	Nature (London) 188: 774- (1960)
Pueraria mirifica (Leguminosae)	Naturwissenschaften 28: 532- (1940)
Pueraria mirifica (Leguminosae)	Indian J Physiol Pharmaco 29 1
Pueraria tuberosa (Leguminosae)	Farmacia (Bucharest) 22: 499-50
Pueraria tuberosa (Leguminosae)	Abstr 27th Annual meeting Ameri
Pueraria tuberosa (Leguminosae)	Planta Med 35: 370-373 (1979)
Pueraria tuberosa (Leguminosae)	Can J Zool 58: 1575-1581 (1980)
Pueraria tuberosa (Leguminosae)	Dokl Bolg Akad Nauk 34: 287-29
Pueraria tuberosa (Leguminosae)	Curr Med Pract 26: 108-111 (19
Pueraria tuberosa (Leguminosae)	Planta Med 1984 2: 154-157 (198
Pueraria tuberosa (Leguminosae)	Vet Sci 13: 59-65 (1976 (Russi
Pueraria tuberosa (Leguminosae)	Bull Med Ethnobot Res 1: 281-2
Pueraria tuberosa (Leguminosae)	Acta Eur Fertil 16 1:59-65 (19
Pueraria tuberosa (Leguminosae)	Indian J Exp Biol 25 7: 442-444
Pueraria tuberosa (Leguminosae)	Planta Med 1986 6:552-B (1986)
Pueraria tuberosa (Leguminosae)	Abstr 11th Conference of Science
Pueraria tuberosa (Leguminosae)	Planta Med 1986 3: 231-233 (198
Punica granatum (Punicaceae) seed oil	J Endocrinol 29: 91- (1964)
Punica granatum (Punicaceae) dried seed	Comp Physiol Ecol 8 2: 101-104
Ranunculus acris (Ranunculaceae)	Can J Anim Sci 60: 531-534 (19
Rheum rhaponticum (Polygonaceae)	Endocrinology 14: 389- (1930) English
Ricinus communis (Euphorbiaceae) seed	Curr Sci 14: 69- (1945) English
Ricinus communis (Euphorbiaceae) seed	Vitamins and Hormones 1954: 207- (1954)
Rubia cordifolia (Rubiaceae)	Indian J Pharmacol 12: 57-58
Salix babylonica (Salicaceae)	Rev Espan Fisiol 16 1: 7- (1960)
Salix caprea (Salicaceae)	Biochem z 180: 1- (1927)
Salix caprea (Salicaceae)	Vitamins and Hormones 1954: 207- (1954)
Salix caprea (Salicaceae)	Vitamins and Hormones 1954: 207- (1954)
Salix species sex female (Salicaceae)	Endocrinologie 1: 411-418 (192
Salvia officinalis (Labiatae)	C R Seances Soc Biol Ses Fil 130: 570- (1939)
Saraca indica (Leguminosae)	Int Conf Chem Biotechnol Biol A
Serenoa repens (Palmae) fruit	Experientia 25: 828- (1969) English
Sesamum indica (Leguminosae)	Riv Zootec Vet 1976: 247- (197
Setaria ciliolata (Gramineae)	Rhod Zambia Malawi J Agr Res 5: 179- (1967)
Solanum dulcamara (Solanaceae)	Rev Espan Fisiol 16 1: 7- (1960)
Solanum tuberosum (Solanaceae) tuber	Vitamins and Hormones 1954: 207- (1954)
Solanum tuberosum (Solanaceae) tuber	J Sci Ind Res-C 15: 202-204 (1
Sphaerophysa salsula (Leguminosae)	Farmakol Prir Veschestv 1978:
Striga orobanchioides (Scrophulariaceae)	Cienc Invest Agr 1: 117-120

Symplocos racemosa (Symplocaceae)	Int Conf Chem Biotechnol Biol A
Syzygium cumini (Myrtaceae)	Indian J Physiol Allied Sci 20: 6- (1967)
Tilia cordata (Tiliaceae)	Rev Espan Fisiol 16 1: 7- (1960)
Tilia platifora (Tiliaceae)	Rev Espan Fisiol 16 1: 7- (1960)
Tillandsia aloifolia (Bromeliaceae)	J Amer Pharm Ass 41: 453- (1952) English
Tillandsia balbisiana (Bromeliaceae)	J Amer Pharm Ass 41: 453- (1952) English
Tillandsia circinnata (Bromeliaceae)	J Amer Pharm Ass 41: 453- (1952) English
Tillandsia fasciculata (Bromeliaceae)	J Amer Pharm Ass 41: 453- (1952) English
Tillandsia juncea (Bromeliaceae)	J Amer Pharm Ass 41:453- (1952) English
Tillandsia simulata (Bromeliaceae)	J Amer Pharm Ass 41:453- (1952) English
Tillandsia tenuifolia (Bromeliaceae)	J Amer Pharm Ass 41:453- (1952) English
Tillandsia usneoides (Bromeliaceae)	J Amer Pharm Ass 41:453- (1952) English
Tinospora cordifolia (Menispermaceae)	Int Conf Chem Biotechnol Biol A
Tribulus terrestris (Zygophyllaceae)	Indian J Pharmacol 13 1: 60
Trifolium alexandrinum (Leguminosae)	Patent-Ger-649,202:- (1937)
Trifolium alexandrinum (Leguminosae)	J Anim Sci 17: 787-791 (1958)
Trifolium fragiferum (Leguminosae)	Aust J Exp Biol Med Sci 27: 297- (1949)
Trifolium fragiferum (Leguminosae)	Arch Biochem Biophys 80: 61- (1959) English
Trifolium fragiferum (Leguminosae)	Science 126: 969- (1957)
Trifolium pratense (Leguminosae)	Experientia 16: 194- (1960) English
Trifolium pratense (Leguminosae)	NZ J Agr Res 4: 328- (1961)
Trifolium pratense (Leguminosae)	S Afr J Sci 59: 561-563 (1963)
Trifolium pratense (Leguminosae)	Indian J Exp Biol 3: 61-63 (19
Trifolium pratense (Leguminosae)	Publ Amer Ass Adv Sci 53: 195- (1959)
Trifolium pratense (Leguminosae)	Tohoku J Exp Med 106: 219- (19
Trifolium pratense (Leguminosae)	Publ Amer Ass Adv Sci 53: 195- (1959)
Trifolium pratense (Leguminosae)	Biochem Z 291: 259-262 (1937)
Trifolium pratense (Leguminosae)	Indian J Pharmacol 12: 60a- (1
Trifolium pratense (Leguminosae)	Nord Vet Med 30:132-136 (1978
Trifolium pratense (Leguminosae)	Patent-Ger-517,761: (1926)
Trifolium pratense (Leguminosae)	Poznan Towarz Przyjaciol Nauk, W
Trifolium pratense (Leguminosae)	Arch Biochem Biophys 80: 61- (1959) English
Trifolium pratense (Leguminosae)	Rast Resur 7: 295- (1971)
Trifolium pratense (Leguminosae)	Biochem Z 291:263-265 (1937)

Trifolium pratense (Leguminosae)	Proc West Sect Amer Soc Anim Sc
Trifolium pratense (Leguminosae)	Contraception 48 2:178-191 (19
Trifolium pratense cv. La Salle (Leguminosae)	Ostrovsky, D: Kitts Wd
Trifolium repens (Leguminosae)	Maslo-Zhir Prom-St 38 1: 26- (1
Trifolium repens (Leguminosae)	Proc XI Ann Conf Indian Pharmac
Trifolium repens (Leguminosae)	Arch Biochem Biophys 80: 61- (1959) English
Trifolium repens (Leguminosae)	Publ Amer Ass Adv Sci 53: 195- (1959)
Trifolium repens (Leguminosae)	Publ Amer Ass Adv Sci 53: 195- (1959)
Trifolium repens (Leguminosae)	Proc Soc Exp Biol 105: 428- (1960)
Trifolium repens (Leguminosae)	Diss Abstr Int B 23: 2286- (1963) English
Trifolium repens (Leguminosae)	Science 126: 969- (1957)
Trifolium repens (Leguminosae)	J Reprod Fertil 6: 115- (1963)
Trifolium repens (Leguminosae)	Poznan Towarz Przyjaciol Nauk w
Trifolium repens (Leguminosae)	Rast Resur 20 3:403-408 (1984)
Trifolium subterraneum (Leguminosae)	J Reprod Fertil 28:160- (1972)
Trifolium subterraneum (Leguminosae)	Bull Haffkine Inst 3: 346, (1
Trifolium subterraneum (Leguminosae)	Aust Vet J 55: 22-24 (1979) En
Trifolium subterraneum (Leguminosae)	J Agr Food Chem 28: 667-671 (1
Trifolium subterraneum (Leguminosae)	Endocrinology 73: 311-313 (195
Trifolium subterraneum (Leguminosae)	Aust J Agr Res 18: 755-765 (19
Trifolium subterraneum (Leguminosae)	Mahidol Univ Annual Res Abstr 1
Trifolium subterraneum cv. Dinninup (Leguminosae)	J Amer Pharm Ass Sci Ed 41:32
Trifolium subterraneum cv. Dwalganup (Leguminosae)	Aust J Exp Biol Med Sci 29: 273- (1951)
Trifolium subterraneum cv. Dwalganup (Leguminosae)	Aust J Exp Biol Med Sci 28: 449- (1950)
Trifolium subterraneum cv. Dwalganup (Leguminosae)	Aust J Exp Biol Med Sci 26:171- (1948)
Trifolium subterraneum cv. Dwalganup (Leguminosae)	Aust J Exp Biol Med Sci 27: 297- (1949)
Trifolium subterraneum cv. Dwalganup (Leguminosae)	J Amer Pharm Ass Sci Ed 41: 32
Trifolium subterraneum cv. Dwalganup (Leguminosae)	Aust Vet J 24: 289-294 (1948)
Trifolium subterraneum cv. Mount Barker (Leguminosae)	Contraception 48 2:178-191 (19
Trifolium subterraneum cv. Tallarook (Leguminosae)	Contraception 48 2:178-191 (19
Trifolium subterraneum cv. Woogeneliup (Leguminosae)	J Amer Pharm Ass Sci Ed 41: 32
Trifolium subterraneum cv. Yarloop (Leguminosae)	Aust J Agr Res 19: 1059-1068 (
Trifolium subterraneum Var. Dwalganua (Leguminosae)	Arch Biochem Biophys 80: 61- (1959) English
Triticum aestivum (Gramineae) seed	C R Seances Soc Biol Ses Fil 157: 1024- (1963)
Triticum aestivum (Gramineae) seed	Bull Acad Natl Med (Paris) 145: 598- (1961)
Triticum aestivum (Gramineae) seed oil	Science 131:1807- (1960)
Triticum aestivum (Gramineae) seed oil	Endocrinology 49: 289- (1951) English
Triticum aestivum (Gramineae) sprouts	J Mammalogy 47: 596- (1966)
Triticum aestivum (Gramineae) dried seed	Comp Physiol Ecol 8 2:101-104
Tulipa gesneriana (Liliaceae)	Arb Med Fakultat Okayama 6: 44
Ulugbekia tschimganica (Boraginaceae)	Farmakol Prir Veschestv 1978:
Vicia americana (Leguminosae)	Mie Daigaku Nogakubu Gakujutsu

Vicia faba (Leguminosae) seed	Planta Med 11:450- (1963)
Vigna radiata (Leguminosae)	Philippine Agr 46: 673- (1963)
Vitex negundo (Verbenaceae) dried seed	Planta Med 1986 6: 552-B (1986)
Yucca aloifolia (Agavaceae)	Rev Espan Fisiol 16 1: 7- (1960)
Zea mays (Gramineae)	Rast Resur 7: 295- (1971)
Zea mays (Gramineae) seed	Biol Pharm Bull 17 8:1029-1031

Appendix A1

Biological activities for compounds from plants with estrogenic activity compiled from University of Illinois at Chicago natural products alert database.

Compound names are listed along with an abbreviated citation. Full citations available from the author upon request.

Acacetin (Flavonoid)	Contraception 42 (4): 467-477 (1990)
Adlupulone (Alicyclic)	Physiology of Reproduction,
Allosponin,alpha (structure unknown)	Rast Resur 12 0:515-525 (1976)
Anethole (Phenylpropanoid)	Biochem J 32641-645 (1938)
Anethole (Phenylpropanoid)	J Pharm Sci 64 (5):717-754 (1975)
Anethole,Di: (Lignan)	Nature (London) 141 78- (1938)
Angolensin (Lignan)	J Endocrinol 52 299- (1972)
Angolensin (Lignan)	J Reprod Fertil 28 (158- (1972)
Anisole (Benzenoid)	J Pharm Sci 64 (5):717-754 (1975)
Anol (Phenylpropanoid)	Nature (London) 139 0:627-628 (1937)
Anol,Di: (Lignan)	Nature (London) 141 78- (1938)
Anol,Di: (Lignan)	Nature (London) 139 (1068- (1937)
Anordrin, alpha (steroid)	Sheng Li Hsueh Pao 36 (6):611-613 (1984)
Anordrin,beta (steroid)	Sheng Li Hsueh Pao 36 (6):611-613 (1984)
Apigenin (Flavonoid)	Mol Pharmacol 44 (L):37-43 (1993)
Apigenin (Flavonoid)	Proc Endocrine Soc 60th Ann Mtg
Apigenin (Flavonoid)	J Pharmacol Exp Ther 224 (2):404 407 (1983)
Asiaticoside (Triterpene)	C R Seances Soc Biol Ses Fil 152 (1110-1114 (1958)
Asiaticoside (Triterpene)	J Pharm Sci 64 (5):717-754 (1975)
Benzofuro-(1'-6':3-4)-coumarin,4'-7-dim (Coumarin)	Arch Biochem Biophys 88 262-266 (1960)
Benzofuro-(1'-6':3-4)-coumarin,4'hyd (Coumarin)	Arch Biochem Biophys 88 262-266 (1960)
Benzofuro-(1'-6':3-4)-coumarin,7-hyd (Coumarin)	Arch Biochem Biophys 88 262-266 (1960)
Benzofuro-(1'-6':3-4)-coumarin,7-hyd (Coumarin)	Arch Biochem Biophys 88 262-266 (1960)

Benzofuro-(1'-6':3-4)-coumarin,7-me: (Coumarin)	Arch Biochem Biophys 88 (162-266 (1960)
Benzoin,2-hydroxy-alpha (Flavonoid)	Indian J Chem Ser B 17 (182-184 (1979)
Bibenzyl,4-4-dihydroxy-5-me (Benzenoid)	Experientia 37 (1181-1182 (1981)
Biochanin A (Flavonoid)	J Endocrinol 52 (199- (1972)
Biochanin A (Flavonoid)	Ann N Y Acad Sci 61 652- (1955)
Biochanin A (Flavonoid)	Aust J Agr Res 18 335- (1967)
Biochanin A (Flavonoid)	J Agr Food Chem 10 110- (1962)
Biochanin A (Flavonoid)	Naturwissenschaften 58 (2):98, (1971)
Biochanin A (Flavonoid)	J Pharm Sci 64 (5):717-754 (1975)
Biochanin A (Flavonoid)	Endeavour 1976 (G):110- (1976)
Biochanin A (Flavonoid)	J Toxicol Environ Health 4 301-324 (1978)
Biochanin A (Flavonoid)	Can J Anim Sci 60 531-534 (1980)
Biochanin A (Flavonoid)	Feedstuffs 51 54-68 (1979)
Biochanin A (Flavonoid)	Chem Ind (London) 92- (1953)
Biochanin A (Flavonoid)	Science 120 (575-576 (1954)
Biochanin A (Flavonoid)	J Endocrinol 24 341-348 (1962)
Biochanin A (Flavonoid)	J Sci Ind Res-B 13 888-889 (1954)
Biochanin A (Filavonoid)	J Amer Pharm Ass Sci Ed 44 404-408 (1955)
Butea superba	Patent-US-2, 136, 397 (1938)
Butin (Flavonoid)	J Ethnopharmacol 18 (L): 95-101 (1986)
Cafesterol (Diterpene)	J Pharm Sci 64 (5): 717-754 (1975)
Cafesterol (Diterpene)	Ber Dtsch Chem Ges 710:1991-1994 (1938)
Cannabidiol (Monoterpene)	J Pharmacol Exp Ther 224 (2):404-407 (1983)
Cannabinol,delta-9-tetrahydro: (Monoterpene)	Science 192 559- (1976)
Cannabinol,delta-9-tetrahydro: (Monoterpene)	Das, Sk: Paria BC: Andrews, Gk: Dey, Sk:
Cannabinol,delta-9-tetrahydro: (Monoterpene)	Science 195 875- (1977)
Cannabinol,delta-9-tetrahydro: (Monoterpene)	Science 195 905-906 (1977)
Capillarin (Coumarin)	Rast Resur 12 515-525 (1976)
Caviuniniso 6-prenyl: (Flavonoid)	Phytochemistry 29 (12): 3921-3922 (1990)
Centchroman (Oxygen heterocycle)	Nigam Pk: Kamboj Vp: Chandra, H
Chalcone 2-4-4'-6-tetrahydroxy: (Flavonoid)	Mol Pharmacol 44 (1): 37-43 (1993)
Chalcone 4-4'-dihydroxy: (Flavonoid)	Mol Pharmacol 44 (1): 37-43 (1993)
Chaplamine (Alkaloid – misc)	Rast Resur 12 515-525 (1976)
Chaplofiligine (Alkaloid – misc)	Rast Resur 12 515-525 (1976)
Chimganin (Sesquiterpene)	Rast Resur 12 515-525 (1976)
Chromene, 2-(H) (Oxygen heterocycle)	J Med Chem 18 982-985 (1975)
Chromene, 2-(H) (Oxygen heterocycle)	J Med Chem 18 982-985 (1975)
Chromene, 2-(H) (Oxygen heterocycle)	J Med Chem 18 982-985 (1975)
Chromene, 2-(H) (Oxygen heterocycle)	J Med Chem 18 982-985 (1975)
Chromene, 2-(H) (Oxygen heterocycle)	J Med Chem 18 982-985 (1975)
Chromene, 2-(H) (Oxygen heterocycle)	J Med Chem 18 982-985 (1975)
Clupanodonic acid (Lipid)	J Pharm Sci 64 (5): 717-754 (1975)
Colchicine (Isoquinoline alkaloid)	J Pharm Sci 64 (5) 117-754 (1975)

Colupulone (Alicyclic)	Physiology of Reproduction
Coronaridine (Indole alkaloid)	J Pharm Sci 62 A199- (1973)
Coronaridine (Indole alkaloid)	Planta Med 33 345-349 (1978)
Coumarin, 3-phenyl-4-7- (Coumarin)	Arch Biochem Biophys 88 162-266 (1960)
Coumestrol (Coumarin)	J Exp Zool 160 319- (1965)
Coumestrol (Coumarin)	J Endocrinol 52 199- (1972)
Coumestrol (Coumarin)	Science 126 969- (1957)
Coumestrol (Coumarin)	J Reprod Fertil 29 A- (1972)
Coumestrol (Coumarin)	J Pharm Sci 53 A411- (1964)
Coumestrol (Coumarin)	Aust J Agr Res 18 335- (1967)
Coumestrol (Coumarin)	J Agr Food Chem 10 10- (1962)
Coumestrol (Coumarin)	J Reprod Fertil 28 (158- (1972)
Coumestrol (Coumarin)	J Pharm Sci 64 (5):717-754 (1975)
Coumestrol (Coumarin)	Endeavour 1976 (G):110- (1976)
Coumestrol (Coumarin)	Biol Reprod 48 (Sl):1 15, (1993)
Coumestrol (Coumarin)	J Toxicol Environ Health 4 301-324 (1978)
Coumestrol (Coumarin)	Biochemistry 16 2896- (1977)
Coumestrol (Coumarin)	Endocrinology 103 860- (1978)
Coumestrol (Coumarin)	Folman, Y Pope, GS:
Coumestrol (Coumarin)	Rast Resur 12 515-525 (1976)
Coumestrol (Coumarin)	Anim Reprod Sci 3 133-245 (1980)
Coumestrol (Coumarin)	Can J Anim Sci 60 53-58 (1980)
Coumestrol (Coumarin)	J Endocrinol 85 317-325 (1980)
Coumestrol (Coumarin)	Can J Anim Sci 60 531-534 (1980)
Coumestrol (Coumarin)	Proc Soc Exp Biol Med 167 (237-241 (1981)
Coumestrol (Coumarin)	Proc Amer Ass Cancer Res 75th Ann Mtg
Coumestrol (Coumarin)	Biochemistry 23 (12): 2565-2572 (191%)
Coumestrol (Coumarin)	J Toxicol Environ Health 15 (L): 51-61 (1985)
Coumestrol (Coumarin)	Steroids 57 (3): 98-106 (1992)
Coumestrol (Coumarin)	Reproductive Toxicology 4 (2)127-135 (1990)
Coumestrol (Coumarin)	Endocrinology 73 736-739 (1963)
Coumestrol (Coumarin)	Arch Biochem Biophys 88 162-266 (1960)
Coumestrol (Coumarin)	J Anim Sci 18 1000-1008 (1959)
Coumestrol (Coumarin)	Nature (London) 218 181-182 (1968)
Coumestrol (Coumarin)	Tech Bull 1408 Aars, USDA, 1408 A-95 (196
Coumestrol (Coumarin)	J Reprod Fertil 210:171-175 (1970)
Coumestrol (Coumarin)	Endocrinology 34 215-225 (1966)
Coumestrol (Coumarin)	J Pharm Sci 53 231-261 (1964)
Coumestrol diacetate (Coumarin)	Feedstuffs 51 54-68 (1979)
Coumestrol, 4'-methoxy: (Coumarin)	J Endocrinol 52 (199- (1972)
Cucurbitacin R (Triterpene)	Experientia 28 (12):1203-1205 (1972)
Cyasterone (steroid)	Rast Resur 12 515-525 (1976)
Daidzein (Flavonoid)	J Endocrinol 52 299- (1972)
Daidzein (Flavonoid)	Ann NY Acad Sci 61 652- (1955)
Daidzein (Flavonoid)	J Agr Food Chem 10 10- (1962)

Daidzein (Flavonoid)	J Reprod Fertil 28 58- (1972)
Daidzein (Flavonoid)	Cancer Res 54 (4): 957-961 (1994)
Daidzein (Flavonoid)	J Toxicol Environ Health 4 301-324 (1978)
Daidzein (Flavonoid)	Osterr Apoth-Ztg 35 121-128 (1981)
Daidzein (Flavonoid)	J Agr Food Chem 28 188-196 (1980)
Daidzein (Flavonoid)	Feedstuffs 51 54-68 (1979)
Daidzein (Flavonoid)	Science 120 575-576 (1954)
Daidzein (Flavonoid)	J Endocrinol 24 341-348 (1962)
Daidzein (Flavonoid)	Aust J Agr Res 19 545-553 (1968)
Daidzein (Flavonoid)	J Amer Pharm Ass Sci Ed 44 404-408 (1955)
Daidzin, 6"-o-malonyl: (Flavonoid)	Acs Symp Ser 546 330-339 (1994)
Daturalactone (steroid)	Indian J Exp Biol 16 648- (1978)
Daturalactone (steroid)	Indian J Exp Biol 16 419-421 (1978)
Deserpidine (Indole alkaloid)	J Pharm Sci 64 (5):717-754 (1975)
Diosgenin (Sapogenin)	Indian J Pharmacy 35 35- (1973)
Diosgenin (Sapogenin)	J Pharm Sci 64 (5):717-754 (1975)
Dorema hyrcanum glycoside	Rast Resur 12 515-525 (1976)
Drupacin (oxygen heterocycle)	Rast Resur 12 515-525 (1976)
Drupanin (Benzenoid)	Rast Resur 12 515-525 (1976)
Dubinidine (Quinoline alkaloid)	Farmakol Prir Veschestv 1978 51-56 (1978)
Dubinidine (Quinoline alkaloid)	Rast Resur 12 515-525 (1976)
Embelin (Quinoid)	Indian J Exp Biol 17 1389-1390 (1979)
Embelin (Quinoid)	Prakash, AO: Saxena, V: Chand, GK: Mathur, R:
Embelin (Quinoid)	Contraception 44 (5): 549-557 (1991)
Ensete Superbum	Fertil Steril 21 247-252 (1970)
Enterolactone (Lignan)	Cancer Res 54 (4):957-961 (1994)
Equol (Flavonoid)	Cancer Res 54 (4):957-961 (1994)
Equol (Flavonoid)	J Endocrinol 85 291-297 (1980)
Equol (Flavonoid)	J Endocrinol 103 395-399 (1984)
Equol (Flavonoid)	Toxicon 3 85-88 (1983)
Equol (Flavonoid)	Biol Reprod 31 105-713 (1984)
Equol (Flavonoid)	Aust J Agr Res 19 545-553 (1968)
Estradiol (steroid)	J Endocrinol 52 299- (1972)
Estradiol (steroid)	J Reprod Fertil 29 1- (1972)
Estradiol (steroid)	J Endocrinol 29 91- (1964)
Estradiol, 17-beta: (steroid)	J Endocrinol 85 291-297 (1980)
Estriol (steroid)	J Endocrinol 52 299- (1972)
Estriol (steroid)	Nature (London) 1310: 766- (1933)
Estrone (steroid)	J Endocrinol 52 299- (1972)
Estrone (steroid)	J Reprod Fertil 29 01- (1972)
Estrone (steroid)	Phytochemistry 5 1337- (1966)
Estrone (steroid)	J Toxicol Environ Health 4 301-324 (1978)
Estrone (steroid)	Feedstuffs 51 54-68 (1979)
Estrone methyl ether (steroid)	J Endocrinol 52 299- (1972)
Estrone, 17-deoxy: (steroid)	J Endocrinol 52 299 (1972)
Eudesmin (Lignan)	Dokl Akad Nauk Uzb Ssr 32 (1): 34- (1975)

Eudesmin (Lignan)	Rast Resur 12 515-525 (1976)
Fagarine, gamma: (Quinoline alkaloid)	Dokl Akad Nauk Uzb Ssr 1982 (9): 34-36 (1982)
Feralin (Sesquiterpene)	Rast Resur 12 515-525 (1976)
Ferujol (Coumarin)	Planta Med 1985 (3) 268-270 (1985)
Flavanone, 4'-7-dihydroxy: (Flavonoid)	Mol Pharmacol 44 (1): 37-43 (1993)
Flavone, 3'-5-7-trihydroxy: (Flavonoid)	Plant Med Phytother 18 (2):74-79 (1984)
Flavone, 4'-6-dihydroxy: (Flavonoid)	J Amer Pharm Ass Sci Ed 45 367- (1956)
Flavone, iso: 4'-me (Flavonoid)	J Nat Prod 53 (1): 62-65 (1990)
Flavone, iso: 7-hydroxy: (Flavonoid)	Patent-Ger Offen-3,415,394 (Aa2 (1983)
Flavone, iso: 7-hydroxy: (Flavonoid)	Patent-Ger Offen-3,430,799 (13pp, (1983)
Flavone, iso: 7-iso-propyl-oxy: (Flavonoid)	Patent-Ger Offen-3,415,394 (Aa2 (1983)
Foliosidine (Quinoline alkaloid)	Rast Resur 12 515-525 (1976)
Formononetin (Flavonoid)	J Chem Soc 1951 3447- (1951)
Formononetin (Flavonoid)	Publ Amer Ass Adv Sci 53 (195- (1959)
Formononetin (Flavonoid)	Science 118 (164- (1953)
Formononetin (Flavonoid)	J Endocrinol 52 (299- (1972)
Formononetin (Flavonoid)	J Endocrinol 52 2w (1972)
Formononetin (Flavonoid)	J Dairy Sci 39 81- (1955)
Formononetin (Flavonoid)	Science 126 969- (1957)
Formononetin (Flavonoid)	J Endocrinol 13 94- (1955)
Formononetin (Flavonoid)	Proc Soc Exp Biol Med 84 506- (1953)
Formononetin (Flavonoid)	Ann N Y Acad Sci 61 652- (1955)
Formononetin (Flavonoid)	J Reprod Fertil 29 (1972)
Formononetin (Flavonoid)	Aust J Agr Res 18 335- (1987)
Formononetin (Flavonoid)	J Agr Food Chem 10 410- (1982)
Formononetin (Flavonoid)	J Agr Food Chem 10 410- (1962)
Formononetin (Flavonoid)	J Reprod Fertil 28 158- (1972)
Formononetin (Flavonoid)	Endeavour 1976 (9) 110- (1976)
Formononetin (Flavonoid)	Endeavour 1976 (9) 110- (1978)
Formononetin (Flavonoid)	Mol Pharmacol 44 (1): 37-43 (1993)
Formononetin (Flavonoid)	J Toxicol Environ Health 4 301-324 (1978)
Formononetin (Flavonoid)	J Toxicol Environ Health 4 301-324 (1978)
Formononetin (Flavonoid)	J Pharm Pharmacol Suppl 37 139-.(1985)
Formononetin (Flavonoid)	Folman, Y: Pope, GS:
Formononetin (Flavonoid)	Anim Reprod Sci 3 233-245 (1980)
Formononetin (Flavonoid)	Osterr Apoth-Ztg 35 121-128 (1981)
Formononetin (Flavonoid)	S-Kh Biol 16 589-592 (1981)
Formononetin (Flavonoid)	J Endocrinol 85 317-325 (1980)
Formononetin (Flavonoid)	Can J Anim Sci 60 531-534 (1980)
Formononetin (Flavonoid)	Can J Anim Sci 60 531-534 (1980)
Formononetin (Flavonoid)	Feedstuffs 510:54-68 (1979)
Formononetin (Flavonoid)	Feedstuffs 51 54-68 (1979)
Formononetin (Flavonoid)	Planta Med 1985 (4): 316-319 (1985)

Formononetin (Flavonoid)	Food Chem Toxicol 23 (8):741-745 (1985)
Formononetin (Flavonoid)	Reproduction & Contraception 2 (2):23-27 (1984)
Formononetin (Flavonoid)	J Pharm Pharmacol Suppl 41 18pp, (1989)
Formononetin (Flavonoid)	Endocrinology 73 736-739 (1963)
Formononetin (Flavonoid)	Science 120 575-576 (1954)
Formononetin (Flavonoid)	Science 120 575-576 (1954)
Formononetin (Flavonoid)	J Endocrinol 24 341-348 (1962)
Formononetin (Flavonoid)	Aust J Agr Res 19 545-553 (1968)
Formononetin (Flavonoid)	Aust J Agr Res 19 545-553 (1968)
Formononetin (Flavonoid)	J Amer Pharm Ass Sci Ed 44 404-408 (1955)
Formononetin (Flavonoid)	J Amer Pharm Ass Sci Ed 44 404-408 (1955)
Formononetin (Flavonoid)	J Reprod Fertil 21 171-175 (1970)
Formononetin (Flavonoid)	Endocrinology 34 215-225 (1966)
Formononetin (Flavonoid)	J Endocrinol 44 567-578 (1969)
Genistin (Flavonoid)	Publ Amer Ass Adv Sci 53 195- (1959)
Genistin (Flavonoid)	Science 118 164- (1953)
Genistin (Flavonoid)	Ann N Y Acad Sci 61 652- (1955)
Genistin (Flavonoid)	J Pharm Sci 64 (5):717-754 (1975)
Genistin (Flavonoid)	Nippon Chikusan Gakkaiho 49 270-275 (1978)
Genistin (Flavonoid)	Food Chem Toxicol 23 (8):741-745 (1985)
Genistin, 6"-o-malonyl: (Flavonoid)	Acs Symp Ser 546 330-339 (1994)
Gibberellic acid (Diterpene)	J Pharm Sci 64 (5):717-754 (1975)
Gibberellin A-3 (Diterpene)	Life Sci 20 785-788 (1977)
Glabranin (Flavonoid)	Rast Resur 12 515-525 (1976)
Glycestrone (structure unknown)	Rast Resur 8 490 (1972)
Glycitin, 6"-o-malonyl: (Flavonoid)	Acs Symp Ser-546 330-339 (1994)
Glycoperine (Quinoline alkaloid)	Doxl Akad Nauk Uzb Ssr 1982 (9):34-36 (1982)
Glycyrrhetinic acid (Triterpene)	Rast Resur 12 515-525 (1976)
Glycyrrhetinic acid (Triterpene)	Egypt J Pharm Sci 16 (245-251 (1975)
Gossypol, (+): (Sesquiterpene)	Indian J Pharmacol 13 (1): 86- (1981)
Guaiacin, iso: 3'-dimethoxy-6-o-demethyl: (Lignan)	J Nat Prod 53 (2):396-406 (1990)
Guaiaretic acid, nor-dihydro: (Lignan)	Cancer Res 54 (4):957-961 (1994)
Halopine (Quinoline alkaloid)	Dokl- Akad Nauk Uzb Ssr 1982 (9):34-36 (1982)
Haplamine (Quinoline alkaloid)	Dokl- Akad Nauk Uzb Ssr 1975 (8):36, (1975)
Jaeschkeanadiol, 5-alpha- (Sesquiterpene)	Planta Med 54 (6): 492-494 (1988)
Jaeschkeanadiol, 5-alpha- (Sesquiterpene)	Planta Med 54 (6): 492-494 (1988)
Jaeschkeanodiol, 5-alpha- (Sesquiterpene)	Planta Med 54 (6): 492-494 (1988)
Kaempferol (Flavonoid)	Cancer Res 54 (4): 957-961 (1994)
Kaempferol (Flavonoid)	Reproduction & Contraception 2 (2):23-27 (1984)
Kopeoside (Coumarin)	Rast Resur 12 515-525 (1976)

Lanatoside c (Cardenolide)	Doxl Bolg Akad Nauk 34 (187-290 (1981)
Liquiritigenin, iso: (Flavonoid)	Mol Pharmacol 44 (1): 37-43 (1993)
Lupulone (Alicyclic)	Physiology of Reproduction
Lupulone, pre: (Alicyclic)	Amer Perf Aromat 75 (5):61-62 (1960)
Luteolin (Flavonoid)	Contraception 42 (4):467-477 (1990)
Mirificin (structure unknown)	Proc Pacific Sce Congr Pacific Sce Assoc 9th
Miroestrol (Flavonoid)	Nature (London) 188 774- (1960)
Miroestrol (Flavonoid)	Naturwissenschaften 28 532-(1940)
Miroestrol (Flavonoid)	J Endocrinol 52 299- (1972)
Miroestrol (Flavonoid)	Proc Pacific Sce Congr Pacific Sce Assoc 9th
Miroestrol (Flavonoid)	Patent-Us-2,112,712 2pp- (1938)
Miroestrol (Flavonoid)	Naturwissenschaften 28 533- (1940)
Miroestrol (Flavonoid)	Terenius L
Miroestrol (Flavonoid)	J Endocrinol 20 (229-235 (1960)
Naringenin (Flavonoid)	Mol Pharmacol 44 (1):37-43 (1993)
Naringin (Flavonoid)	S Afr J Lab Clin Med 4 289- (1958)
Nicotine (Alkaloid)	J Pharm Sce 64 (5):717-754 (1975)
Pent-4-enoic acid (Benzenoid)	Vet Arh Suppl 510:111-113 (1981)
Perforine (Quinoline alkaloid)	Farmakol Prier Veschestv 1978 51-56 (1978)
Perforine (Quinoline alkaloid)	Rast Resur 12 515-525 (1976)
Phaseollin (Flavonoid)	Biochem Z 291 263-265 (1937)
Phloretin (Flavonoid)	Proc Soc Exp Biol Med 114 115- (1963)
Phloretin (Flavonoid)	J Pharm Sce 64 (5):717-754 (1975)
Phloretin (Flavonoid)	Mol Pharmacol 44 (1):37-43 (1993)
Pilocarpine (Alkaloid)	J Pharm Sce 64 (5):717-754 (1975)
Podocarpic acid (Diterpene)	J Pharm Sce 64 (5):717-754 (1975)
Podocarpinol (Diterpene)	Nature (London) 161 892- (1948)
Podocarpinol, 7-iso-propyl: (Diterpene)	J Amer Chem Soc 72 3800- (1950)
Prunetin (Flavonoid)	Pharmacol Plant Phenolics
Prunetin (Flavonoid)	J Pharm Sce 6.4 (5):717-754 (1975)
Psoralen (Coumarin)	Rast Resur 12 515-525 (1976)
Puerarin (Flavonoid)	Proc Pacific Sce Congr Pacific Sce Assoc 9th
Quercetin (Flavonoid)	S Afr J Lab Clin Med 4 289- (1958)
Reserpine (Indole alkaloid)	J Pharm Sce 64 (5):717-754 (1975)
Rhaponticin (Benzenoid)	Die Medizinische 1956 195- (1956)
Rhaponticin (Benzenoid)	J Pharm Sce 64 (5):717-754 (1975)
Robinin (Flavonoid)	S Afr J Lab Clin Med 4 289- (1958)
Scoparone (Coumarin)	Rast Resur 12 515-525 (1976)
Segetalin A (Proteid)	Tetrahedron 51 (2l):6003-6014 (1995)
Segetalin B (Proteid)	Tetrahedron 51 (2l):6003-6014 (1995)
Sitosterol, beta: (steroid)	Zentralbl Veterinaermed 1 1a(A76-(1964)
Sitosterol, beta: (steroid)	Pharmazie 25 620 (1970)
Sitosterol, beta: (steroid)	Experimentia 25 828- (1969)

Sitosterol, beta: (steroid)	Z Naturforsch Ser B 20 0:817- (1965)
Sitosterol, beta: (steroid)	Naturwissenschaften 51 (A09- (1964)
Sitosterol, beta: (steroid)	Planta Med 19 0:208- (1971)
Skimmianine (Quinoline alkaloid)	Rast Resur 12 515-525 (1976)
Sophoricoside (Flavonoid)	Kiserl Orvostud 13 (133- (1961)
Soyasaponin B-1 (Triterpene)	Acs Symp Ser 546 330-339 (1994)
Soyasaponin B-B (Triterpene)	Acs Symp Ser 546 330-339 (1994)
Stigmasterol (steroid)	Pharmazie 25 620 (1970)
Stigmasterol (steroid)	Ann Biochem Exp Med 21 141- (1961)
Stilbene (Quinoid)	J Endocrinol 2 444- (1941)
Stilbene, 4-4'-dihydroxy: (Benzenoid)	J Endocrinol 2 444- (1941)
Stilbene, 4-hydroxy: (Benzenoid)	J Endocrinol, 2 444- (1941)
Stilbene, trans: (Benzenoid)	Yao Hsueh Hsueh Pao 14 (4)227-233 (1979)
Tectorigenin (Flavonoid)	Nippon Yakurigaku Zasshi 64 186- (1968)
Tricin (Flavonoid)	J Pharm Sci 53 (1411- (1964)
Turkesterone (steroid)	Rast Resur 12 515-525 (1976)
Tyrosine (Proteid)	Sheng Chih Yu Bi Yun 2 (1)23-27 (1982)
Veratraldehyde (Benzenoid)	Biochem J 32641-645 (1938)
Vincamine (Indole alkaloid)	Dokl Akad Nauk Uzb Ssr 31 (3):33- (1974)
Vitamin D-2 (Vitamin)	IRCS Med Sci Libr Compend 8 (286- (1980)
Viticosterone E (steroid)	Rast Resur 12 515-525 (1976)
Wax (Ananas comosus) (Lipid)	Science 121 42- (1955)
Wogonin (Flavonoid)	Phytother Res 4 (3):86-89 (1990)
Yohimbine (Indole alkaloid)	Obstet Ginecol 2 390-395 (1945)
Yuehchukene (Indole alkaloid)	Eur J Pharmacol 264 11-12 (1994)
Yuehchukene (Indole alkaloid)	Yao Hsueh Hsueh Pao 25 (2):85-89 (1990)
Yuehchukene (Indole alkaloid)	Eur J Med Chem 26 (4):387-394 (1991)

Appendix B

Biological activities for compounds from plants with antispermatogenic acitivity compiled from University of Illinois at Chicago natural products alert database

Plant scientific names are listed along with an abbreviated citation. Full citations available from the author upon request.

Benzofurano-(6-7-B) (Coumarin)	Naturwissenschaften 67 104- (1980)
Berberine (Isoquinoline alkaloid)	Fitoterapia 61 (1): 67-71 (1990)
Berberis chitria alkaloid (Alkaloid misc)	India Today 1981 (5): 98- (1981)
Caffeine (Alkaloid)	Food Chem Toxicol 22 (5): 365-369 (1984)
Caffeine (Alkaloid)	J Environ Pathol Toxicol 2 687-706 (1979)
Caffeine (Alkaloid)	J Reprod Fertil 63 11-15 (1981)
Cannabidiol (Monoterpene)	Adv Biosci Marihuana Bio Eff 22 407-418 (1978)
Cannabidiol (Monoterpene)	Science 216 315-316 (1982)
Cannabinol (Monoteripene)	Adv Biosci Marihuana Bio Eff 22 407-418 (1978)
Cannabinol (Monoterpene)	Science 216 315-316 (1982)
Cannabinol, delta-9-tetrahydro: (Monoterpene)	Adv Biosci Marihuana Bio Eff 22 407-418 (1978)
Cannabinol, delta-9-tetrahydro: (Monoterpene)	Surg Forum 27 350-352 (1976)
Cannabinol, delta-9-tetrahydro: (Monoterpene)	Science 216 315-316 (1982)
Cathinone, (-) (Isoquinoline alkaloid)	Toxicology 60 (3) 223-234 (1990)
Cathinone, (+) (Isoquinoline alkaloid)	Toxicology 60 (3) 223-234 (1990)
Colchicine (Isoquinoline alkaloid)	Pharm Urinary Tract Male Reprod Syst 1982 0:325
Colchicine (Isoquinoline alkaloid)	Fertil Steril 23 180-181 (1972)
Colchicine (Isoquinoline alkaloid)	Pharmacol Rev 11 0:135-172 (1959)
Colebrookia flavonoid (Flavonoid)	India Today 1981 (5): 98- (1981)
Cyclohexanol (Alicyclic)	Indian J Exp Biol 17 0: 1305-1307 (1979)
Demecolcine (Isoquinoline alkaloid)	Arch Ostet Ginecol 69 327- (1963)
Dopa, L (Proteid)	Indian J Med Res 71 46-52 (1980)
Embelin (Ouinoid)	Andrologia 15 (5):486-494 (1983)
Embelin (Quinoid)	Androl 18 (2):125-131 (1986)
Flavone, 2-hydroxy (Flavonoid)	Contraceptive Delivery Systems
Flavone, 2-hydroxy- (Flavonoid)	Arch Androl 9 (1): 28-. (1982)
Flavone, 3'-5-7-trihydroxy (Flavonoid)	Plant Med Phytother 20 (2): 188-198 (1986)
Flavylium, 3-4-7-trihydroxy (Flavonoid)	Contraceptive Delivery Systems
Flavylium ,3-4-7-trihydroxy-t (Flavonoid)	Arch Androl 9 (11) 18, (1962)
Glycyrrhetinic acid, alpha: (Triterpene)	Arch Sci Med 1068 836-847 (1958)
Gossypol (Sesquiterpene)	J Med 22 (1): 29-44 (1991)
Gossypol (Sesquiterpene)	Verh Anat Ges 79 (2):571-572 (1985)
Gossypol (Sesquiterpene)	Int J Androl 11 (6) 533-546 (1988)

Gossypol (Sesquiterpene)	Chung-Hua I Hsueh Tsa Chih 4 417-428 (1978)
Gossypol (Sesquiterpene)	Lancet 1979 (1): 395- (1979)
Gossypol (Sesquiterpene)	Acta Biol Exper Sin 11 15-22 (1978)
Gossypol (Sesquiterpene)	Newsweek Feb (25): 9-10 (1980)
Gossypol (Sesquiterpene)	Yao Hsueh Hsueh Pao 14 663-669 (1979)
Gossypol (Sesquiterpene)	Contraception 21 461-469 (1980)
Gossypol (Sesquiterpene)	Gynecol Obstet Invest 10 163-176 (1979)
Gossypol (Sesquiterpene)	Abstr Soc Study Reprod Aug 11-14 149- (1980)
Gossypol (Sesquiterpene)	Recent Advances in Fert Reg. 122-146 (1981)
Gossypol (Sesquiterpene)	Cell Tissue Res 204 293-296 (1979)
Gossypol (Sesquiterpene)	Acta Acad Med Wuhan 1979 102-105 (1979)
Gossypol (Sesquiterpene)	Prog In Hormone Biochem Pharma: 149-225 (1980)
Gossypol (Sesquiterpene)	Natl Med J China 58 455-458 (1978)
Gossypol (Sesquiterpene)	Proc Who-Nih Meeting Nov 17-18 1980 7pp- (1980)
Gossypol (Sesquiterpene)	Chieh P'ou Hsueh Pao 11 299-302 (1980)
Gossypol (Sesquiterpene)	Res Frontiers In Fertility Regulation 1 (4) 1-15 (1981)
Gossypol (Sesquiterpene)	Wo Hsueh Tung Pao 25 1049-1050 (1980)
Gossypol (Sesquiterpene)	Compar Med East West 8 195-197 (1980)
Gossypol (Sesquiterpene)	Rung Wu Hsueh Pao 26 311-316 (1980)
Gossypol (Sesquiterpene)	Clin Pharmacology & Therapeutics 489-492 (1981)
Gossypol (Sesquiterpene)	Nieschlag, E: Wickings, EJ: Breuer, H
Gossypol (Sesquiterpene)	14th Ann Soc For Study Of Reprod 14 111A- (1981)
Gossypol (Sesquiterpene)	14th Ann Soc For Study Of Reprod 14 35A- (1981)
Gossypol (Sesquiterpene)	Int J Androl Suppl 5 53-70 (1982)
Gossypol (Sesquiterpene)	Natl Med J China 59 402-405 (1979)
Gossypol (Sesquiterpene)	Shih Yen Sheng Wu Hsueh Pao 13 193-200 (1980)
Gossypol (Sesquiterpene)	Osterr Apoth-Ztg 35 121-128 (1981)
Gossypol (Sesquiterpene)	Int J Androl 3 507-518 (1980)
Gossypol (Sesquiterpene)	Andrologia 13 242-249 (1981)
Gossypol (Sesquiterpene)	Yao Hsueh T'ung Pao 15 (9):42- (1980)
Gossypol (Sesquiterpene)	T'ung Wu Hsueh Pao 27 (22-28 (1981)

Gossypol (Sesquiterpene)	Contraception 24 97-105 (1981)
Gossypol (Sesquiterpene)	Amer Pharm Ns21 (11):57-59 (1981)
Gossypol (Sesquiterpene)	Chung-Kuo Yao Li Hsueh Pao 2 (262-266 (1981)
Gossypol (Sesquiterpene)	Cell Tissue Res 219 659-663 (1981)
Gossypol (Sesquiterpene)	Recent Advances In Fertility Regulation
Gossypol (Sesquiterpene)	Recent Advances In Fertility Regulation
Gossypol (Sesquiterpene)	Recent Advances In Fertility Regulation
Gossypol (Sesquiterpene)	N Engl J Med 303 334-336 (1980)
Gossypol (Sesquiterpene)	J Androl 2 190-199 (1981)
Gossypol (Sesquiterpene)	Yao Hsueh Hsueh Pao 17 14 (1982)
Gossypol (Sesquiterpene)	Recent Advances In Fertility Regulation
Gossypol (Sesquiterpene)	IRCS Med Sci 10 766-769 (1982)
Gossypol (Sesquiterpene)	Ann NY Acad Sci 383 458-459 (1982)
Gossypol (Sesquiterpene)	Biol Reprod 27 241-252 (1982)
Gossypol (Sesquiterpene)	Nippon Funin Gakkai Zasshi 26 393-397 (1981)
Gossypol (Sesquiterpene)	Chieh P'ou Hsueh Pao 13 211-214 (1982)
Gossypol (Sesquiterpene)	Chieh P'ou Hsueh Pao 13 101-205 (1982)
Gossypol (Sesquiterpene)	Isr J Med Sci 17 742 (1981)
Gossypol (Sesquiterpene)	Arch Androl 8 233-246 (1982)
Gossypol (Sesquiterpene)	Tohoku J Exp Med 138 275-280 (1982)
Gossypol (Sesquiterpene)	Biol Reprod 28 (4): 1007-1020 (1983)
Gossypol (Sesquiterpene)	Contraception 27 (4): 391-400 (1983)
Gossypol (Sesquiterpene)	Med Res Rev 2 403-432 (1982)
Gossypol (Sesquiterpene)	Int J Fertil 27 213-218 (1982)
Gossypol (Sesquiterpene)	Sheng Chih Yu Bi Yun 2 (2): 52-55 (1982)
Gossypol (Sesquiterpene)	Patent-Us-4.381, 298 6pp, (1983)
Gossypol (Sesquiterpene)	Arch Androl 9 (1) 37-38 (1982)
Gossypol (Sesquiterpene)	Biol Reprod 26 183-195 (1982)
Gossypol (Sesquiterpene)	Arch Androl 9 (1) 39-43 (1982)
Gossypol (Sesquiterpene)	Arch Androl 9 (1) 35- (1982)
Gossypol (Sesquiterpene)	Arch Androl 9 (1) 32-33 (1982)
Gossypol (Sesquiterpene)	J Reprod Fertil 70 (1)341-345 (1984)
Gossypol (Sesquiterpene)	Urol Res 11 (2):75-85 (1983)
Gossypol (Sesquiterpene)	Chung-Kuo Yao Li Hsueh Pao 4 (3): 183-185 (1983)
Gossypol (Sesquiterpene)	Arch Androl 9 (1) 36- (1982)
Gossypol (Sesquiterpene)	Hsueh Yuan Hsueh Pao 5 (4): 227-
Gossypol (Sesquiterpene)	Biol Reprod 30 (5) 1198-1207 (1984)
Gossypol (Sesquiterpene)	J Reprod Fertil 71 (1): 127-133 (1984)
Gossypol (Sesquiterpene)	Rev Pure Appl Pharmacol Sci 3 1: 1-81 (1982)
Gossypol (Sesquiterpene)	Fertil Steril 42 (3): 424-430 (1984)

Gossypol (Sesquiterpene)	Contraceptive Delivery Systems 5 (3):10-11 (1984)
Gossypol (Sesquiterpene)	Contraceptive Delivery Systems 5 (3):14-15 (1984)
Gossypol (Sesquiterpene)	Endocrine Mechanisms in Fertility Regulation 1983
Gossypol (Sesquiterpene)	Internl Sympon Chinese Medicinal
Gossypol (Sesquiterpene)	J Androl 5 (6): 416-423 (1984)
Gossypol (Sesquiterpene)	Chung-Kuo Yao Li Hsueh Pao 3 (4): 260-263 (1982)
Gossypol (Sesquiterpene)	Lancet (1984) (II): 107- (1984)
Gossypol (Sesquiterpene)	Kalla, NR: Foo, JTW: Hurkadli, KS: Sheth, AR:
Gossypol (Sesquiterpene)	Yao Hsueh Hsueh Pao 20 (5): 392-394 (1985)
Gossypol (Sesquiterpene)	Handbook of Natural Toxins 1 785-815 (1983)
Gossypol (Sesquiterpene)	Contraception 32 (6): 651-660 (1985)
Gossypol (Sesquiterpene)	Econ & Medicinal Plant Research Vol 1
Gossypol (Sesquiterpene)	Curr Top Reprod Endocrinol 2: 145-157 (1982)
Gossypol (Sesquiterpene)	Adv in Chinese Med Mat Res
Gossypol (Sesquiterpene)	J Androl 6 (2):Abstr-G3 (1985)
Gossypol (Sesquiterpene)	Adv in Chinese Med Mat Res
Gossypol (Sesquiterpene)	J Androl 6 (2):Abstr-M12 (1985)
Gossypol (Sesquiterpene)	J Androl 6 (2):Abstr-M3 (1985)
Gossypol (Sesquiterpene)	Potential Contraceptive for Men
Gossypol (Sesquiterpene)	Int J Androl 8 (3):177-185 (1985)
Gossypol (Sesquiterpene)	Contraceptive Delivery Syst 5 (4): 335-356 (1984)
Gossypol (Sesquiterpene)	Int Congr Ser-Excerpta Med 655 227-230 (1984)
Gossypol (Sesquiterpene)	Andrology J 3 (7):127-139 (1986)
Gossypol (Sesquiterpene)	Ontogenez 16 (4):346-357 (1985)
Gossypol (Sesquiterpene)	Acta Histochem 77 (2):185-191 (1985)
Gossypol (Sesquiterpene)	J Ethnopharmacol 20 (1): 21-24 (1987)
Gossypol (Sesquiterpene)	Bull WHO 65 (4):551-552 (1987)
Gossypol (Sesquiterpene)	Bull WHO 65 (4):547-548 (1987)
Gossypol (Sesquiterpene)	Contraception 36 (5): 581-592 (1987)
Gossypol (Sesquiterpene)	Contraception 37 (2): 137-151(1988)
Gossypol (Sesquiterpene)	Contraception 37 (2): 129-135 (1988)
Gossypol (Sesquiterpene)	Contraception 37 (2): 153-162 (1988)
Gossypol (Sesquiterpene)	Adv Contraceptive Delivery Syst 3 183-193 (1987)
Gossypol (Sesquiterpene)	Contraceptive Delivery Systems
Gossypol (Sesquiterpene)	Int J Androl 11 (1):1-11 (1988)
Gossypol (Sesquiterpene)	Arch Androl 9 (1): 28- (1982)
Gossypol (Sesquiterpene)	Int J Androl 10 (4): 619-623 (1987)

Gossypol (Sesquiterpene)	Chung-Kuo I Hsueh K'o Hsueh Yuan Hsueh Pao 12 (6):440
Gossypol (Sesquiterpene)	Indian J Med Res 87 368-376 (1988)
Gossypol (Sesquiterpene)	Cancer 35 (4): 395-408 (1987)
Gossypol (Sesaulterpene)	Wang, JM: Qui, JP: Wu, Xl: Zhang, ZR: Shao,Y: Cao, RQ:
Gossypol (Sesquiterpene)	Fertil Steril 48 (3): 462-465 (1987)
Gossypol (Sesquiterpene)	Chin J Integ Trad West Med 8 (1): 36, (1988)
Gossypol (Sesquiterpene)	Int J Androl 13 (4) 253-257 (1990)
Gossypol (Sesquiterpene)	Drugs of the Future 17 (5): 421-422 (1992)
Gossypol (Sesquiterpene)	Acta Biol Exper Sin 11 27-30 (1978)
Gossypol (Sesquiterpene)	Proc Nat Conf on Recent Adv
Gossypol (Sesquiterpene)	Proc Nat Conf on Recent Adv
Gossypol (Sesquiterpene)	Proc Nat Conf on Recent Adv
Gossypol (Sesquiterpene)	Personal Communication (1978)
Gossypol (Sesquiterpene)	Unpublished data through WHO 8pp- (1980)
Gossypol (Sesquiterpene)	Personal communication from WHO (1979)
Gossypol, (+) (Sesquiterpene)	J Ethnopharmacol 20 (1): 65-78 (1987)
Gossypol, (DL) (Sesquiterpene)	Biol Reprod 37 (4) 924 (1987)
Harringtonine (Isoquinoline alkaloid)	Chung-Hua I Hsueh Tsa Chih 92 175-180 (1979)
Hesperidin methyl chalcone (Flavonoid)	Science 118 657- (1953)
Hippadine (Isoquinoline alkaloid)	Planta Med 49 (4) 252-254 (1983)
Hydroquinone, meta-xylo: (Benzenoid)	J Med Int Med Abstr 23 33-(1959)
Hydroquinone, meta-xylo: (Benzenoid)	J Med Int Med Abstr 22 19-25 (1958)
Hydroquinone, meta-xylo: (Benzenoid)	Sci Cult 25 661-665 (1960)
Hydroquinone, meta-xylo: (Benzenoid)	Bull Calcutta Sch Trop Med 10 (2): 85-89 (1962)
Kaempferol (Flavonoid)	Plant Med Phytother 23 (3): 193-201 (1989)
Keracyanin (Flavonoid)	Nara Igaw Zasshi 30 (1) 13-19 (1979)
Leurocristine (Indole alkaloid)	J Pharmacol Exp Ther 181 12- (1972)
Leurocristine (Indole alkaloid)	Curr Sci 68 (10):1053-1057 (1995)
Leurocristine (Indole alkaloid)	Cancer Treat Rep 63 35- (1979)
Leurocristine (Indole alkaloid)	Exp Pathol 15 85- (1978)
Leurocristine (Indole alkaloid)	Brit J Pharmacol 63 677-681 (1978)
Leurocristine (Indole alkaloid)	Pharma Urinary Tract Male Reprod Syst 1982 325
Leurocristine (Indole alkaloid)	Arch Toxicol Suppl 7 151-154 (1984)
Leurocristine (Indole alkaloid)	Curr Sci 63 (3): 144-147 (1992)
Lycorine (Isoquinoline alkaloid)	Pak J Sci Ind Res 4 280- (1961)
Malvidin (Flavonoid)	Curr Sci 57 (24) 1354-1355 (1988)
Malvidin (Flavonoid)	Int J Androl 13 (3) 207-215 (1990)
Nicotine (Alkaloid)	Irc Med Sci 4 519- (1976)

Nicotine (Alkaloid)	Andrologla 11 273-278 (1979)
Oleanolic acid (Triterpene)	J Ethnopharmacol 24 (1): 115-121 (1988)
Palmatine (Isoquinoline alkaloid)	J Ethnopharmacol 25 (2): 151-157 (1989)
Palmatine, dihydro: (Isoquinoline alkaloid)	Contraceptive Delivery Systems
Palmatine, dihydro: (Isoquinoline alkaloid)	Arch Androl 9 (1) 28 (1982)
Palmatine, dihydro: (Isoquinoline alkaloid)	Fitoterapia 61 (1): 67-71 (1990)
Phthalate, di-(2-ethyl-hexyl): (Benzenoid)	Environ Health Perspect 45 77-84 (1982)
Phthalate, di-n-butyl: (Benzenoid)	Environ Health Perspect 45 77-84 (1982)
Phthalate, di-n-hexyl: (Benzenoid)	Environ Health Perspect 45 77-84 (1982)
Phthalate, di-n-pentyl: (Benzenoid)	Environ Health Perspect 45 77-84 (1982)
Phthalate, di-n-pentyl: (Benzenoid)	J Pathol 139 (3) 309-321 (1984)
Phthalate, mono-(2-ethyl-hexyl): (Benzenoid)	Environ Health Perspect 45 77-84 (1982)
Phthalate, mono-iso-butyl: (Benzenoid)	Environ Health Perspect 45 77-84 (1982)
Phthalate, mono-n-butyl: (Benzenoid)	Environ Health Perspect 45 77-84 (1982)
Phthalate, mono-n-pentyl: (Benzenoid)	Environ Health Perspect 45 77-84 (1982)
Phthalate, mono-sec-butyl: (Benzenoid)	Environ Health Perspect 45 77-84 (1982)
Plumbagin (Quinoid)	Indian J Pharmacol 13 (1): 66-. (1981)
Plumbagin (Quinoid)	Indian J Exp Biol 22 153-156 (1984)
Preparation roc-101 (structure unknown)	Indian J Physiol Pharmacol 16 181- (1972)
Preparation roc-101 (structure unknown)	Indian J Med Res 60 1213-1219 (1972)
Protopine (Isoquinoline alkaloid)	Fitoterapia 61 (1): 67-71 (1990)
Reserpine (Indole alkaloid)	Wiss Z Paedagog Hochsch 22 59-64 (1978)
Retinoic acid, all-trans: (Diterpene)	J Androl 6 (2) Abstr-M54 (1985)
Retinol acetate (Vitamin)	Iyaklihin Kenkyu 12 1064-1081 (1981)
Sitosterol, beta: (steroid)	Libyan J Sci Ser B 8 17-24 (1978)
Sitosterol, beta: (steroid)	J Ethnopharmacol 35 (2) 149-153 (1991)
Solasodine (Steroid alkaloid)	Dixit, VP: Gupta, RS:
Syringic acid (Benzenoid)	Indian J Appl Chem 28 132-140 (1965)
Syringic acid (Benzenoid)	J Med Int Med Abstr 28 97- (1964)
Taxifolin (Flavonoid)	Science 118 657- (1953)
Taxol (Diterpene)	J Toxicol Sci 19 (S-1): 11-34 (1994)
Theobromine (Alkaloid)	Toxicon 70 (2) 155-164 (1994)
Theobromine (Alkaloid)	Food Chem Toxicol 22 (5) 265-369 (1984)
Theobromine (Alkaloid)	J Environ Pathol Toxicol 2 687-706 (1979)

Theophylline (Alkaloid)	J Environ Pathol Toxicol 2 687-706 (1979)
Tripchlorolide (Diterpene)	Proc Chin Acad Med Sci 4 (3): 17
Tripchlorolide (Diterpene)	Hsueh Yuan Hsueh Pao 12 (6): 440
Tripchlorolide (Diterpene)	Patent-Faming Zhuanli Shenqing Gongkai
Tripchlorolide (Diterpene)	Jiepou Xuebao 24 (2) 204-207 (1993)
Tripdiolide (Diterpene)	Contraception 47 (4): 387-400 (1993)
Tripterygium T-2-1 (structure unknown)	Proc Chin Acad Med Sci 4 (3) 17
Tripterygium T-2 (structure unknown)	Proc Chin Acad Med Sci 4 (3) 17
Triptolide, 15-hydroxy: (Diterpene)	Patent-Faming Zhuanli Shenqing Gongkai
Tryptophan, L: (Proteid)	Andrologia 14 142-249 (1982)
Vincaleukoblastine (Indole alkaloid)	Exp Cell Res 76 470- (1973)
Vincaleukoblastine (Indole alkaloid)	Proc Serono Symp Male Fert Steril 5 423- (1974)
Vincaleukoblastine (Indole alkaloid)	Exp Pathol 15 85- (1978)
Vincaleukoblastine (Indole alkaloid)	Brit J Pharmacol 63 677-681 (1978)
Vincaleukoblastine (Indole alkaloid)	Pharma Urinary Tract Male Reprod Syst 1982 325
Vincaleukoblastine (Indole alkaloid)	Arch Biol 85 353-364 (1974)
Vincaleukoblastine (Indole alkaloid)	Male Ferrility And Sterility
Vindesine	Cancer Treat Rep 63 35- (1979)

Appendix B1

Biological activities for extracts of plants with antispermatogenic activity compiled from University of Illinois at Chicago natural products alert database

Plant scientific name listed along with an abbreviated citation. Full citations available from the author upon request.

Abrus precatorius (Leguminosae)	J Ethnopharmacol 28 2: 173-181 (1990) English
Abrus precatorius (Leguminosae)	Geobios 8 1: 29-31 (1981) English
Allium sativum (Liliaceae) dried bulb	Indian J Exp Biol 20: 534-536 (1982) English
Andrographis paniculata (Acanthaceae)	Indian J Exp Biol 28 5: 421-426 (1990) English
Apium graveolens (Umbellifereae)	J Med Ass Thailand 64 4: 164-172 (1979)
Aristolochia indica (Aristolochiaceae)	Indian J Exp Biol 15: 256- (1977)
Azadirachta indica (Meliaceae)	J Androl 14 4: 275-281 (1993) English
Azadirachta indica (Meliaceae)	Curr Sci 64 9:688-689 (1993) English

Balanites roxburghii (Zygophyllaceae)	Indian J Exp Biol 19: 918-921 (1981) English
Bambusa arundinacea (Gramineae)	J Ethnopharmacol 25 2:173-180 (1989) English
Butea monosperma (Leguminosae)	Jugoslav Physiol Pharmacol Acta 17: 151-162 (1981)
Butea monosperma (Leguminosae)	Jugoslav Physiol Pharmacol Acta 17: 151-162 (1981)
Butea monosperma (Leguminosae)	J Sci Res Pl Med 5 112: 27-30 (1984) English
Cannabis sativa (Cannabaceae)	Proc Nat Conf on Advances in Fam Plan Res (1972)
Catharanthus roseus (Apocynaceae)	Indian J Exp Biol 6: 256-257 (1968)
Catharanthus roseus (Apocynaceae)	Indian J Exp Biol 29 9: 810-812 (1991) English
Celastrus paniculatus (Celastraceae)	J Androl 6 2: Abstr-N18 (1985) English
Celastrus paniculatus (Celastraceae)	J Ethnopharmacol 28 3: 293-303 (1990) English
Celastrus paniculatus (Celastraceae)	Fitoterapia 59 5: 377-382 (1988) English
Cichorium intybus (Compositae)	Naturwissenschaften 70 7:365-366 (1983) English
Ecballium elaterium (Cucurbitaceae)	Patent-Us-4, 148, 892 (1979)
Embelia ribes (Myrsinaceae)	Indian J Exp Biol 17: 935-938 (1979) English
Gelsemium sempervirens (Loganiaceae)	J Amer Inst Homeopathy 19: 707-713 (1926) English
Gossypium hirsutum (Malvaceae)	Acta Acad Med Wuhan 1979: 102-105 (1979) German
Gossypium hirsutum (Malvaceae)	Acta Biol Exper Sin 11: 1-10 (1978) Chinese
Gossypium hirsutum (Malvaceae)	Proc WHO-Nih Meeting Chemical Synthesis
Gossypium hirsutum (Malvaceae)	Yao Hsueh Hsueh Pao 14: 663-669 (1979) Chinese
Gossypium hirsutum (Malvaceae)	J Nutr 112: 2052-2057 (1982) English
Helianthus annuus (Compositae)	Zhivotnovodstvo 32 1: 78- (1970)
Hibiscus rosa-sinensis (Malvaceae)	J Reprod Fertil 38: 233-234 (1974) English
Hibiscus rosa-sinensis (Malvaceae)	J Res Indian Med 7: 72-73 (1972) English
Hibiscus rosa-sinensis (Malvaceae)	Eur J Pharmacol 26:111-114 (1974) English
Hibiscus rosa-sinensis (Malvaceae)	J Res Indian Med Yoga Homeopathy 9 4:99-102 (1974)
Hippophae salicifolia (Elaeagnaceae)	Indian J Exp Biol 3: 206-208 (1985) English
Lupinus albus (Leguminosae)	Aust J Biol Sci 22: 1071- (1969) English
Madhuca latifolia (Sapotaceae)	Food Chem Toxicol 28 9:601-605 (1990) English

Malvaviscus conzattii (Malvaceae)	Indian Counc Med Res-Ann 1978: 63-64 (1978)
Malvaviscus conzattii (Malvaceae)	Proc Indian Nat Acad Sci Ser B 44: 1-6 (1978)
Malvaviscus conzattii (Malvaceae)	Contraception 43 3: 273-285 (1991) English
Malvaviscus conzattii (Malvaceae)	Contraception 31 1: 101-108 (1985) English
Malvaviscus conzattii (Malvaceae)	Indian J Exp Biol 18: 561-564 (1980) English

10 The saga of the falling sperm counts

James Le Fanu

Summary

Recently, attention has been drawn to the assertion that chemical pollutants which mimic the effect of the female hormone, oestrogen, known as xenoestrogens, are linked to low sperm counts, infertility and genital abnormalities.

At the same time, published scientific reports show an increase in sperm count. Other published research rules out any possibility of xenoestrogens impairing fertility or sexual function. These findings have been considered little, or not at all, in several reports on the effects of such chemicals on reproductive health.

This paper reviews the literature, and finds a reliance on one piece of work – a meta-analysis of many separate studies, made over many years, in different countries – as evidence of a decline in sperm count. Flaws in this analysis make any assumption of falling sperm counts unsafe.

An environmental threat to fertility?

The saga of falling sperm counts may seem only one of many serious environmental threats to health brought to the public's attention over the last decade, but it is particularly worthy of study because its implications are truly apocalyptic. Traffic pollution may or may not cause asthma, nitrate fertilizers may or may not cause stomach cancer, but homo sapiens will survive such misfortunes (see Chapter 6). If, however, it is true that the chemical pollutants which mimic the effect of the female hormone, oestrogen – known as xenoestrogens – are compromising the potential of males to propagate the species, then clearly, 'something must be done', and fast.

Two recent events have highlighted the need to evaluate the validity of current concerns. The first popular account of the saga of falling sperm counts was published in May 1996 – *Our Stolen Future*, by Theo Colborn, John Patterson Myers and Dianne Dumanoski. Its theme is well summarised in an introduction by no less than the Vice President of the United States, Al Gore: 'Emerging scientific research [shows] how a wide range of man-made chemicals disrupt delicate hormone systems which play a crucial role in human sexual development and have been linked to low sperm counts, infertility and genital deformities.'

Almost simultaneously, however, the scientific journal, *Fertility and Sterility*, reported two separate studies, drawing on data from the United States, which show that there has been absolutely no change in sperm counts over the past few decades (Lipshulz 1996). How, one might rightly ask, can a widely-acclaimed book, with an introduction by the Vice President of the United States, which highlights a serious threat to the survival of the human species, appear concurrently with studies which show that no such threat exists? Clearly they can't both be right.

Probably the best way of answering this question is to follow the story as it has unfolded over the past few years and which, as will be seen, provides interesting lessons on the nature and origins of current concerns about environmental threats to health. The starting point was a paper published in the *British Medical Journal* in September 1992 by researchers in Denmark (Carlsen and Skakkebaek 1992). They had reviewed 61 scientific papers, published between 1938 and 1990, covering data on nearly 15 000 men, which revealed a 'highly significant' decrease in mean sperm count from 113×10^6 per ml in 1940 to 66×10^6 per ml in 1990. This they linked with a two-fold rise in incidence of testicular cancer over this period, and suggested that a common explanation for the two phenomena might be that 'oestrogens or oestrogen-like compounds' might in some way 'be damaging testicular function'.

Six months later, in May 1993, Professor Skakkebaek, along with Dr Richard Sharpe, of the Medical Research Council's Reproduction Biology Unit in Edinburgh, published a further paper in *The Lancet* in which they elaborated on the evidence implicating oestrogens in the decline of sperm counts (Sharpe and Skakkebaek 1993). They started by drawing attention to the results of work published almost a decade earlier, which showed 'a substantial increase' in abnormalities of the male genital tract and 'decrease in sperm counts' in men born to mothers who had been given the drug, diethylistilboestrol – DES – during pregnancy, which had been widely prescribed in the fifties for the prevention of miscarriage (Stillman 1982).

They then speculated that similar effects on the male foetus, during a vulnerable stage of its development, may have been brought about by maternal exposure to xenoestrogenic chemical pollutants: 'Many of the chemicals with which we have contaminated our environment in the past 50 years are weakly oestrogenic.' They concluded that: 'The most reasonable (and safest) assumption is that pregnant women are exposed to more, rather than less, oestrogens than was the case 50 years ago.'

By now, and inevitably, Professor Skakkebaek's thesis had attracted the attention of the media, and in October of the same year, the prestigious BBC science programme, *Horizon*, produced 'Assault on the Male' which investigated the claims that falling sperm counts were caused by oestrogenic chemicals in meat and water.

In 1995, the campaigning organisation, Greenpeace, entered the fray with a national advertising campaign to highlight the issue. Under the eye-catching slogan, 'You're not half the man your father was', the text went on: 'Its true. Scientists have shown that the same chemicals we dump in our seas are causing willies to shrink.' To drive the point home there was an accompanying picture of a boy with what appeared to be a very small willy – taken from a fifteenth century fresco. This alleged decline in willy size was linked with data reported by Professor Skakkebaek of a secular decline in sperm counts and rise in male genital abnormalities and testicular cancer.

The claims made by Greenpeace received apparently authoritative scientific backing in June 1995, with the publication of a report from the Medical Research Council's Institute for Environmental Health, at Leicester University (Institute for Environmental Health 1995). This reviewed all the appropriate data, and though concluding that there was no definitive proof that xenoestrogens were threatening male fertility, nonetheless expressed the opinion that there was sufficient uncertainty to warrant further research. Coming from such an eminent source, this report naturally generated enormous media interest. The *Daily Mail* headlined the story: 'Scientists fear everyday products may spell doom for mankind', and the *Daily Telegraph*: 'Gender-bending chemicals may cause cancer'.

In February 1996, Dr Stuart Irvine, also of the Medical Research Council's Reproductive Biology unit in Scotland, reported in the *British Medical Journal*, a study of cohorts of men born between 1951 and 1973, where the mean sperm concentration had fallen from 98×10^6 per ml to 78×10^6 per ml (Irvine 1996). An accompanying editorial from Dr D. M. de Kretser, Director of the Institute of Reproduction and Development at Monash University in Australia, noted that 'oestrogens and pesticides are implicated' and warned that: 'Delay in instituting further studies to confirm or refute these hypotheses, may compromise the fertility of future generations' (de Kretser 1996).

So, in just under four years, Professor Skakkebaek's original paper had been taken up and amplified, a plausible mechanism in xenoestrogenic chemicals had been identified, and the public's attention had been drawn to the importance of the issue by the media and a brilliant Greenpeace propaganda campaign. Where, if anywhere, was the catch?

It is necessary first to get some idea of the scale of the problem that has to be dealt with by establishing the quantity of the xenoestrogenic pollutants to which the human species is exposed. In Jonathan Tolman's contribution to this publication – 'Nature's hormone factory' – he makes it clear that there is a vast number of 'naturally-occurring' substances with oestrogenic properties. Some of these compounds – the bioflavonoids – are an intrinsic part of the human diet and one can presume that the human organism must have adaptive mechanisms that prevent damage to the male reproductive tract from their consumption. The important question then, is how much the xenoestrogens from chemical pollutants contribute to our intake of oestrogen over and above that which we might naturally consume.

Dr Stephen Safe, of the Department of Veterinary Physiology and Pharmacology at Texas A&M University, has provided the answer. His findings were published in the journal, *Environmental Health Perspectives*, in April 1995 (Safe 1995). He found that: 'The dietary intake of xenoestrogens is 0.0000025 per cent of the daily intake of oestrogenic flavonoids in the diet'. He concluded that 'their contribution to male reproductive problems is not plausible'.

This is a crucial observation for any evaluation of the potential dangers of xenoestrogens, but the Medical Research Council's publication already alluded to – which generated the 'Gender-bending chemicals may cause cancer' headlines – dealt with it in just one paragraph tucked away almost halfway through the report (Institute for Environmental Health 1995). Dr de Kretser's authoritative editorial in the *British Medical Journal* of February 1996, 'Declining sperm counts – environmental chemicals may be to blame', deals with Dr Safe's findings by simply not mentioning them (de Kretser 1996).

This question of the scale of exposure also arises when evaluating the significance of the findings in men who have been exposed to the drug DES, while in utero. It will be recalled that Professor Skakkebaek and Dr Richard Sharpe had cited evidence of low sperm counts and abnormalities of the male genital tract in these men as evidence that, by analogy, the exposure of pregnant women to xenoestrogens might also explain a worldwide decline in sperm count (Sharpe and Skakkebaek 1993).

There are, however, two limitations to this interpretation. First, it is by no means clear that DES did indeed have the effect that they claim. The most recent evaluation from the Mayo Clinic has failed to identify any increase in 'genito-urinary abnormalities, infertility or testicular cancer' in men who were exposed to DES in utero (Leary 1984). Further, Dr Allen J. Wilcox and colleagues from the University of Virginia, in a paper in the *New England Journal of Medicine* in 1995, reported that: 'High doses of DES did not lead to impairment of fertility or sexual function in adult men who had been exposed to the drug in utero' (Wilcox 1995). Even if, for the sake of argument, DES exposure had had an adverse effect on male reproductive capacity, the scale of exposure is of a different order of magnitude from that of xenoestrogen to which pregnant women may or may not be exposed. The dosage of DES ranged from 700 mg up to 12 600 mg throughout pregnancy. So, with an average of 6000 mg, this works out at around 20 mg per day. This is billions of times greater than a pregnant woman and her foetus would be exposed to from xenoestrogenic pollutants.

So, if concentration of xenoestrogens in the environment is virtually immeasurably small, and the analogy with men exposed to DES in utero no longer holds, then there can only be two possibilities – either there is another explanation for declining sperm counts or, the decline itself is a myth.

This takes us back to the first publication, back in 1992, that started the hare running: Professor Skakkebaek's analysis of the 61 scientific studies over the previous 50 years. The problems with this publication are three-fold. First, over this period the official criterion for a 'normal' sperm count fell markedly. So, many men who would have been excluded from the earlier studies (because their sperm counts were 'abnormally low') would have been included in the later studies. Thus, by definition, the mean average sperm counts appeared to fall over time (Bromwich 1994). Secondly, Professor Skakkebaek had failed to compare like with like. Studies showing a 'high' sperm count in the 1940s and 1950s were mostly drawn from the United States, while the later ones, showing a 'fall', came from Europe and less developed countries. As there is a well-recognised geographical variation in mean sperm counts, his findings could not be considered to provide any consistent data in changes in sperm counts over time (Lipshulz 1996). Thirdly, it would appear that when the analysis of sperm counts is limited to the last 20 years, it is not possible to show any consistent evidence of decline (Brake and Krauser 1992). One might surmise that the methods of measurement of sperm counts would have become more accurate in recent years, in which case the lack of change over the last 20 years is likely to be more reliable than the evidence of a substantial fall in sperm count from previously 'high' levels measured four or five decades ago.

The current situation can thus be summarised in the following terms. Those who wish to believe that declining sperm counts threaten the survival of the human species may wish to point to the evidence from the Scottish study, published in the *British Medical Journal*. They must, nonetheless, acknowledge that this has not been confirmed by similar studies from other countries, such as those published in *Fertility and Sterility*. They must also acknowledge that the inference from Skakkebaek's original paper in 1992 of a historical decline in sperm counts, cannot be justified on the basis of the evidence that is presented. They must also acknowledge that his inferred mechanism of decline – exposure to xenoestrogens – cannot be sustained for the reasons put forward by Dr Stephen Safe, and that the analogy drawn with men exposed to DES in utero is invalid.

References

Brake, A., Krauser, W. (1992). *British Medical Journal* **305**, 1498.
Bromwich, P. (1994). *British Medical Journal* **309**, 19.
Carlsen, E., Skakkebaek, N. E. (1992). *British Medical Journal*, **305**, 609.
Colborn, T., Myers, J., Dumanoski, D. (1996). *Our Stolen Future*. Little, Brown, Boston.
de Kretser, D. M. (1996). *British Medical Journal* **312**, 457.
Institute for Environmental Health (1995). *Environmental Oestrogens: Consequence to Human Health and Wildlife*, Medical Research Council.
Irvine, S. (1996). *British Medical Journal* **312**, 457.
Leary, F. J. (1984). *Journal of the American Medical Association.* **252**, 2984.
Lipshulz, L. I. (1996). *Fertility and Sterility*, **65**, 909.
Safe, S. (1995). *Environmental Health Perspectives*, **103**, 346.
Sharpe, R. M., Skakkebaek N. E. (1993). *The Lancet* **341**, 1392.
Stillman, R. J. (1982). *American Journal of Obstetrics and Gynaecology* **142**, 905.
Wilcox A. J. (1995). *New England Journal of Medicine*, **332**, 1411.

11 Commentary

Lorraine Mooney

Worrying about small risks is a sign of the times – at least in post-industrial economies. What sociologist, Frank Furedi refers to as the 'anxiety about existence' shows itself in our susceptibility to panic about the long term health of ourselves or our environment.

It is part of man's nature to try to improve his condition. As he grows richer, he will tolerate less hazard and discomfort. Western societies have come so far along this path that the remaining dangers are no longer obvious. We would not even be aware of many hazards that now preoccupy us if science and technology had not developed the means to identify and measure them, and to study their effects.

Such is our sophistication that the identification, management and removal of risks has become the business of governments and international agencies. But, in handing over responsibility for health and safety to an agent, the judgement process of weighing risk against risk – the trade-off – has been impaired. Of course, the information necessary to make that trade-off is too much for any individual to acquire. It is now necessary for an expert in risk management to assess the evidence of other experts, before a judgement can be made.

Unfortunately, cool consideration is often made impossible by clamourers with a single implacable view of what should be done, and who are adept at swaying public opinion. Regulation becomes inevitable as a political expediency rather than as a problem solver, and 'regulators ... are sometimes influenced by the public's present tendency toward chemophobia' and fail to adequately weigh risks and benefits (Malaspina 1992). For instance, US laws which prohibit the import of produce containing chemical pesticide residue, forced Mexican vegetable farmers in the Culiacan Valley to switch from organochlorines to the less persistent organophosphates recommended by the US EPA. Sadly, they are also more acutely toxic than the ones they gave up. The consumers' presumed gain was made at the expense of agricultural workers in a poorer country (Perfecto 1992).

As Kemm has stated, in the case of the DDT campaign, it is not even a question of a human health risk. The risk–risk trade off concerned here is balanced between the possible reproductive impairment of birds of prey and the lives of two million people a year – mainly children. That is not to say that they could all be saved, but the effort could be made.

Now, despite there being stores of DDT all over Africa, (although Western countries stopped using DDT some have continued its manufacture) it now suffers the stigma of being rejected by the more sophisticated Americans. The international health agencies are reluctant to propose using DDT for the sake of political correctness, but there are signs that Zimbabwe is willing to try using it again to control mosquitoes.

Time and science have largely refuted the claims made against DDT, but even if they were true, they are surely trivial compared to the human death toll. The new leadership of the WHO under Gro Harlem Brundtland is making its 'Roll Back Malaria' campaign a priority. At the time of writing details have not been released, but it would be welcome if it was decided to increase the proportion of annual resources spent on it from the current five per cent. WHO are known to recognise the value of DDT, so it would also be interesting to see if they are brave enough to seriously consider using it again.

It is part of the WHO's remit to be normative, to set the standard of how things should be. At the moment, the new leadership policy unit is busy discussing the expansion of the meaning of 'well-being' from the 'absence of disease or infirmity' that was the understanding of the founders fifty years ago, to 'a state of complete physical, mental and social well-being'. Further declaring that the enjoyment of the highest attainable standard of health is one of the fundamental rights of every human being. The consensus definition of these 'rights' amongst WHO officials include 'the right to adequate food, water, clothing, health care, education, reproductive health and social services, and the right to security in case of unemployment, sickness, disability, old age, or lack of livelihood in circumstances beyond an individual's control. Respect for human rights and the achievement of public health goals are complementary' (WHO 1998). This may or may not be helpful, it is difficult to say because it is so hard to understand what it means and how it might work. How to enforce a 'right to adequate clothing', how much education and what type? These statements raise more questions than they are worth, and there is plenty more such rhetoric to philosophise over, including this gem:

> 'Imperialistic, universalistic and essentialist approaches to philosophy, ethics and social policy are gradually being transformed as they come under the growing influences of multiculturalism, feminism, non-racialism and other reactionary responses to the failures of the past... More meaningful global values such as empathy, respect for other cultures, a deeper understanding of human rights coupled with responsibilities will replace the fads sweeping the world markets' (Benatar, 1997).

However 'meaningful' these sentiments are in Geneva, such sentiments are of little value to those not rich enough to protect themselves from communicable diseases.

It is perhaps inevitable that international agencies based in rich, cosmopolitan cities will be more enthused by health problems that are more immediate or novel to their members and funders, and that their policies will reflect their own preferences. Westerners want to save developing countries now from the problems that they might encounter in the future, rather than help them to deal with the problems that they are actually facing today. There have actually been seat-belt campaigns in parts of Africa where the only vehicles for a hundred miles are aid-agency Land Rovers. Perhaps it is boring for WHO officials to be faced year after year with the same intractable problems to deal with, but that is, surely, what is expected of them. The way forward suggested by Dr Adetokunbo Lucas (1998) is that WHO should vary its spending in each country according to its needs, having found that some of the poorest countries received the smallest input of WHO resources. He also suggests that WHO should defer to existing national or international agencies already operating in a country, rather than duplicating efforts. And, tellingly, he finds that the profile of expertise in WHO offices is inconsistent with national requirements.

Now that western society has reaped the benefits of economic development in the form of long and healthy lives and high levels of consumption, it takes them for granted. It even disparages economic growth – a view seemingly endorsed by the WHO, amazingly. We hope that developing countries will not do themselves the disservice of emulating Western attitudes which are irrelevant or dangerous to them. We further hope that Western expertise and assistance is used to help poorer people in a way that is useful to them.

References

Benatar, S. (1997). Key ethical dimensions of the renewal process at the global level: streams of global change, in: *Ethics, Equity and Health for All*, Edited by Z. Bankowski, J. H. Bryant and J. Gallagher, CIOMS.

Lucas, A. (1998). WHO at country level, *The Lancet*, **351**, 743–747.

WHO (1998a) *Health for All in the 21st century*, Document A51/5:2.

About ESEF

Mission Statement

The European Science and Environment Forum is an independent, non-profit-making alliance of scientists whose aim is to ensure that scientific debates are properly aired, and that decisions which are taken, and action that is proposed, are founded on sound scientific principles.

The ESEF will be particularly concerned to address issues where it appears that the public and their representatives, and those in the media, are being given misleading or one-sided advice. In such instances the ESEF will seek to provide a platform for scientists whose views are not being heard, but who have a contribution to make.

Members are accepted from all walks of life and all branches of science. There is no membership fee. Members will be expected to offer their services in contributing to ESEF publications on issues where their expertise is germane.

Purpose of ESEF

The European Science and Environment Forum is a Charitable Company Limited by Guarantee (No. 1060751). It was established in 1994 to inform the public about scientific debates. Our chosen method for achieving this objective is to provide a forum for scientific opinions that are usually not heard in public policy debates.

Our primary role is to provide an independent voice to the media, the general public and the educators, and by doing so, we aim to provide balance on scientific issues. A secondary role is to contribute to the scientific debate itself. Many of our authors will simplify papers that they originally wrote for the peer reviewed scientific literature. ESEF's tertiary role is to advise scientists how to present their findings to the media, and how the media will perceive, and may use, the information. We hope that this will provide dialogue and understanding between these two important institutions.

How was ESEF formed?

ESEF was formed in 1994 by Roger Bate (Director of the Environment Unit at the Institute of Economic Affairs, London), Dr John Emsley (Science Writer in Residence at Imperial College London University) and Professor Frits Böttcher (Director of the Global Institute for the Study of Natural Resources in The Hague). The issue of climate change was the initiation for the meeting. All three thought that the debate had been unduly one-sided and they wanted to provide a forum for scientists to publish their arguments for public consumption. The media, and via them the public, tended to only hear the so-called consensus view presented by government and intergovernmental science panels.

Of course climate change is not the only issue where member scientists consider that the media debate is not balanced and that there are many environmental and public health issues which are not fully discussed in the public arena either.

ESEF decided on a mission to provide the media and the public with accessible first-hand research of leading scientists in their fields, as an alternative to reports received from specialist journals, government departments or single-issue pressure groups. As Einstein is reputed to have said: 'Make science as simple as possible, but no simpler'. Our aim is to provide science simplified as far as possible. Our members are from fields as diverse as nuclear physics, biochemistry, glaciology, toxicology and philosophy of science. We intend on liaising between the media and our expert members to provide an independent voice on subjects germane to various public policy debates.

To maintain its independence and impartiality, ESEF accepts funding only from charities, and the income it receives is from the sale of its publications. Such publications will be sent to selected opinion formers within the media and within government.

Academic Members of ESEF

August 1998

Prof. Tom Addiscott UK
Prof. Bruce Ames USA
Dr Alan Bailey UK
Dr Sallie Baliunas USA
Dr Robert C. Balling USA
Prof. A. G. M. Barrett UK
Dr Jack Barrett UK
Mr Roger Bate UK
Dr Sonja Boehmer-Christiansen UK
Prof. Frits Bottcher The Netherlands
Prof. Norman D. Brown UK
Prof. Dr K. H. Büchel Germany
Dr John Butler UK
Mr Piers Corbyn UK
Prof. Dr. A. Cornelissen The Netherlands
Dr Tom Coultate UK
Dr Barrie Craven UK
Mr Peter Dietze Germany
Dr A. J. Dobbs UK
Dr John Dowding UK
Dr John Emsley UK
Dr Oeystein Faestoe UK
Dr Patricia Fara UK
Dr Frank Fitzgerald UK
Prof Dr Hartmut Frank Germany
Dr James Franklin Belgium
Dr Alastair Gebbie UK
Dr T. R. Gerholm Sweden

Prof. Dr Gerhard Gerlich Germany
Prof. D. T. Gjessing, Norway
Dr Manoucher Golipour UK
Dr Adrian Gordon Australia
Dr Michael Gough USA
Dr Vincent R. Gray New Zealand
Dr Gordon Gribble USA
Prof Dr Hans-Eberhard Heyke Germany
Dr Vidar Hisdal Norway
Dr Jean-Louis L'hirondel France
Dr Sherwood Idso USA
Dr Antoaneta Iotova Bulgaria
Prof. Dr Zbigniew Jaworowski Poland
Dr Tim Jones UK
Prof. Dr Wibjörn Karlén Sweden
Dr Terrence Kealey UK
Prof. Dr Kirill Ya. Kondratyev Russia
Prof. Dr F. Korte Germany
Mr Johan Kuylenstierna Sweden
Dr Theodor Landscheidt Germany
Dr Alan Mann UK
Dr John Marks UK
Dr John Mason UK
Dr John McMullan UK
Prof. Dr Helmut Metzner Germany
Dr Patrick Michaels USA
Sir William Mitchell UK
Dr Paolo Mocarelli Italy
Dr Asmunn Moene Norway
Dr Brooke T. Mossman USA
Prof Dr Hans-Emil Müller Germany
Prof Dr Dr Paul Müller Germany

Dr Joan Munby UK
Mr Liam Nagle UK
Dr Genrik A. Nikolsky Russia
Dr Robert Nilsson Sweden
Prof. Dr Harry Priem The Netherlands
Dr Christoffer Rappe Sweden
Dr Ray Richards UK
Dr Michel Salomon France
Dr Tom V. Segalstad Norway
Dr S. Fred Singer USA
Dr Willie Soon USA
Dr G. N. Stewart UK
Dr Gordon Stewart UK
Dr Maria Tasheva Bulgaria
Dr Wolfgang Thüne Germany
Dr Alan Tillotson UK
Dr Brian Tucker Australia
Prof. Dr med. Karl Überla Germany
Prof. Dr H. P. van Heel The Netherlands
Dr Robin Vaughan UK
Prof. Nico Vlaar The Netherlands
Dr Horst Wachsmuth Switzerland
Dr Michael P. R. Waligórski Poland
Dr Gunnar Walinder Sweden
Dr Gerd-Rainer Weber Germany
Prof Donald Weetman UK
Dr Charlotte Wiin-Christensen Denmark
Dr Aksel Wiin-Nielsen Denmark
Dr James Wilson USA

Business Members

Dr Alfred Bader UK
Mr John Boler UK
Mr Charles Bottoms UK
Dr Francisco Capella Gómez-Acebo Spain
Mr Richard Courtney UK
Dr Wynne Davies UK
Dr Claes Hall UK
Mr Richard Hallett UK
Mr Peter Henry UK
Mr Holger Heuseler Germany
Mr Graham Horne UK
Dr Warwick Hughes Australia
Dr Kelvin Kemm South Africa
Mr Peter Plumley UK
Dr John Rae UK
Dr Michael Rogers UK
Mr Peter Toynbee Australia

Authors' addresses

Roger Bate
ESEF
4 Church Lane
Barton
Cambridge
CG3 7BE
Great Britain
E-mail: roger@esef.org

Dr B. M. Craven
Newcastle Business School
University of Northumbria
Newcastle upon Tyne
NE1 8ST
Great Britain
E-mail: barrie.craven@unn.ac.uk

Dr Enrique Ghersi
eghersi@mail.ccec.edu.pe

Gordon W. Gribble, Ph.D.
Department of Chemistry
Dartmouth College
Hanover, NH 03755-3564
USA
E-mail: Grib@Dartmouth.edu

Dr Frank Furedi
Reader in Sociology
Darwin College
The University of Kent at Canterbury
Kent CT2 7NY
Great Britain
E-mail: F.Furedi@ukc.ac.uk

Christine Johnson
P O Box 2424
Venice,
California
90294-2424
USA
E-mail: ay409@lafn.org

Dr Kelvin Kemm
Stratek: Technology Strategy Consultants
PO Box 74416
Lynnwood Ridge
Pretoria 0040
South Africa
E-mail: stratek@pixie.co.za

Prof. Dr K. Y. Kondratyev
Research Centre for Ecological Safety
Russian Academy of Sciences
Korpusnaya Street, 18
St Petersburg 197110
Russia
E-mail nansen@sovam.com

Dr James Le Fanu
24 Grafton Square
London
SW4
Great Britain

Michael Fumento
4600 North Washington Boulevard, 207
Arlington
Virginia 22201
USA
E-mail mfumento@compuserve.com

Dr Jean-Louis L'Hirondel
Centre Hospitalier Régional et Universitaire de Caen
Service de Rhumatologie
Avenue de la Côte de Nacre
14033 CAEN CEDEX
France

Michelle M. Malkin
The Seattle Times
Seattle,
Washington
USA

Lorraine Mooney
ESEF
4 Church Lane
Barton
Cambridge
CB3 7BE
Great Britain
E-mail: lorraine@esef.org

Hector Ñaupari
libertario@mixmail.com

Jonathan Tolman
Competitive Enterprise Institute
1001 Connecticut Avenue, N.W.
Suite 1250
Washington DC 20036
USA
E-mail: jtolman@cei.org

Index

Agenda 21, 81
Alar, 95
American Public Health Association, 136
Ames, B. 69
at risk concept, 47–59

Beauty Parlour Syndrome, 53
benzene, 67
BSE 52, 88, 93, 100–108

campylobacter, 89, 132
Carson, R. 7, 62, 144, 152
causes of cancer, 69
chlorine 17–25, 30–33, 72, 129–176
cholera 17–42, 51, 131–136, 137, 154, 171
climate warming 63, 65
clover (*Trifolium subterraneum*) 178
Creutzfeldt-Jakob disease, 86, 115

DDT 1–16, 71, 131, 143–163, 177–184
DES 178, 217–221
dioxin, 73, 74, 86, 137–186
dose threshold levels 67

E. coli 87, 122, 171
Environmental Defence Fund (EDF) 10
Environmental Protection Agency (EPA) 6, 13, 17, 131–156
environmental tobacco smoke (ETS), 69
European Commission's Scientific Committee for Food 127

Food and Agricultural Organisation (FAO) 120, 127
Food and Drug Administration (FDA) 146, 186
food law 103
flue gas desulphurisation technology (FGD) 79

Galbraith, J. K. 90
General Household Survey 48
Gough, M. 147, 149, 168
Greenpeace 129–143, 152, 218

hormetic effects 77

International Agency for Research on Cancer 129, 137
International Commission on Radiological Protection (ICRP) 73
International Joint Commission, 140

Jaworowski, Z. 63

Kealey, T. 80

listeria, 52, 87

Ministry of Agriculture, Fisheries and Food (MAFF) 86, 106, 109
malaria 1, 15, 51, 53, 142, 145, 170

Manhattan Project 74
Measles Mumps Rubella (MMR) vaccine 48
meta-analysis 140, 152, 215
metabolism of nitrates 120
methicillin-resistant staphylococcus aureus (MRSA) 52

outrage 81, 96, 103, 113

phytoestrogen 180–183
plant defensive systems 185
Policy Studies Institute, 48
polychlorinated biphenyls (PCBs) 145
precautionary principle 153
PVC 172

radioactive pollution 63
risk perception 99, 103

Safe, S. 150, 184, 218
salmonella 52, 87, 100
Sandman, P. 81, 96
shellfish 20, 30, 34, 89
Shubik, p. 143
Skakkebaek, N. 151, 218–221
soya beans 180–183
sperm count 159, 215–220
stomach cancer 122, 215

The Earth Charter 82
The Food Standards Agency 109
threshold dose level 63–74
Toxic Oil Syndrome 87

United Nations Scientific Committee on the Effects of Atomic Radiation (UNSCEAR) 74

Vahrenholt, F. 80

Water and Sanitation for Health (WASH) 244
Wildavsky, A. 137
World Health Organisation (WHO) 1, 10, 17, 34, 67, 119–132, 149

xenoestrogens 184, 216, 220